TSTI-Waco
3 6705 00024358 6

MEASUREMENT OF AIR POLLUTANTS

Guide to the Selection of Methods

MEASUREMENT OF AIR POLLUTANTS

Guide to the Selection of Methods

M. KATZ, M.Sc., Ph.D.

Professor, Department of Civil Engineering,
L. C. Smith College of Engineering,
Syracuse University, Syracuse, N.Y., USA

WORLD HEALTH ORGANIZATION

GENEVA

1969

The designations employed and the presentation of the material in this publication do not imply the expression of any opinion whatsoever on the part of the Director-General of the World Health Organization concerning the legal status of any country or territory or of its authorities, or concerning the delimitation of its frontiers.

The author alone is responsible for views expressed in this publication.

The mention of specific companies or of certain manufacturers' products does not imply that they are endorsed or recommended by the World Health Organization in preference to others of a similar nature which are not mentioned. Errors and omissions excepted, the names of proprietary products are distinguished by initial capital letters.

PRINTED IN SWITZERLAND

CONTENTS

LIST OF REVIEWERS

Dr A. Bell, Director, Division of Occupational Health, New South Wales Department of Health, Sydney, New South Wales, Australia

Mr J. J. Bloomfield, Institute of Occupational Health and Air Pollution Research, Santiago, Chile

Professor P. Bonnevie, Director, University Institute of Hygiene, Copenhagen, Denmark

Professor P. Chovin, Director, Municipal Laboratory, Paris, France

Dr L. Coin, Head, City Public Health Laboratory, Paris, France

Dr S. R. Craxford, Director, Warren Spring Laboratory, Stevenage, Herts., England

Professor J. P. Détrie, Technical Centre for the Study of Air Pollution, Paris, France

Professor Lars T. Friberg, Institute of Hygiene, Karolinska Institute, Stockholm, Sweden

Professor A. Giovanardi, Director, University Institute of Hygiene, Milan, Italy

Dr L. Greenburg, Chairman, Albert Einstein College of Medicine, New York, N.Y., USA

Mr S. Griswold, Chief, Abatement Branch, National Center for Air Pollution Control, Washington, D.C., USA

Dr E. C. Halliday, National Physical Research Laboratory, Pretoria, South Africa

Dr F. Hartogensis, TNO Institute of Public Health Engineering, The Hague, The Netherlands

Professor T. F. Hatch, School of Public Health, University of Pittsburgh, Pa., USA

Dr E. Heller, Director, Federal Institute for Air, Water and Soil Hygiene, West Berlin, Germany

Professor E. Hettche, Landesanstalt für Immissions u. Bodennutzungschutz, Essen, Federal Republic of Germany

Professor D. Högger, Head, Occupational Health Service, Zurich, Switzerland

Dr K. Horiuchi, Professor of Preventive Medicine and Public Health, Osaka City University, Osaka, Japan

Dr N. F. Izmerov, Deputy Minister of Health and Chief Sanitary Inspector of the Ministry of Health, Moscow, USSR

Professor J. Just, Director, Department of Communal Hygiene, State Institute of Hygiene, Warsaw, Poland

Dr B. Kesić, Director, Andrija Stampar School of Public Health, Zagreb, Yugoslavia

Professor T. Kitagawa, Department of Safety in Chemical Engineering, Yokohama National University, Yokohama, Japan

Mr J. Krämer, Secretary, VDI Kommission " Reinhaltung der Luft ", Düsseldorf, Federal Republic of Germany.

Dr P. J. Lawther, Director, MRC Air Pollution Research Unit, London, England

Professor E. Leclerc, Director-General, Belgian Centre for Air and Water Studies, Liège, Belgium

Mr P. Lieben, Directorate for Scientific Affairs, Organization for Economic Co-operation and Development, Paris, France

Dr J. H. Ludwig, Chief, Laboratory of Engineering and Physical Sciences, Robert A. Taft Sanitary Engineering Center, Cincinnati, Ohio, USA

Mr V. G. MacKenzie, Chief, Division of Air Pollution, US Public Health Service, Washington, D.C., USA

Mr R. S. Mehta, Director, Central Public Health Engineering Research Institute, Nagpur, India

Dr M. H. MERRILL, Director, California State Department of Public Health, Berkeley, Calif., USA
Dr R. MORF, Secretary-General, International Union of Pure and Applied Chemistry
Professor Y. NOSE, Department of Public Health, Yamaguchi Medical University, Ube City, Japan
Professor B. PACCAGNELLA, Director, University Institute of Hygiene, Ferrara, Italy
Professor F. L. PETRILLI, Director, University Institute of Hygiene, Genoa, Italy
Professor V. A. RJAZANOV, Director, Institute of General and Municipal Hygiene, Moscow, USSR
Professor August T. ROSSANO, University of Washington, Seattle, Wash., USA
Professor A. ROUSSELL, National Institute of Health and Medical Research, Paris, France
Professor F. H. SCHMIDT, Royal Netherlands Meteorological Institute, De Bilt, The Netherlands
Professor L. SILVERMAN, Head, Department of Industrial Hygiene, Harvard School of Public Health, Boston, Mass., USA
Professor H. E. STOKINER, Chief, Toxicology Section, Occupational Health and Training Facility, US Public Health Service, Cincinnati, Ohio, USA
Professor K. SYMON, Director, Institute of Hygiene, Prague, Czechoslovakia
Professor J. W. TESCH, TNO Public Health Research Organization, The Hague, The Netherlands
Professor R. TRUHAUT, Faculty of Pharmacy, University of Paris, France
Professor V. B. VOUK, Director, Institute of Medical Research, Zagreb, Yugoslavia
Professor P. W. WEST, College of Chemistry and Physics, Louisiana State University, Baton Rouge, La., USA

Preface

Within recent years, there has been increasing concern over the need for establishing recommended procedures for air sampling and analysis. Although many recommendations have been made by various governmental and nongovernmental agencies, much remains to be done. Agreement is needed on methods for determining the concentration of common pollutants, on methods for determining the effects of pollutants on health and vegetation, on guides for air quality, on nomenclature, and on units of measurement. Only when such agreement has been reached will the results of air pollution studies in different areas and countries be fully comparable.

The importance of uniformity in this field cannot be overemphasized. Air pollution is a growing hazard to health throughout the world, and authorities in both developed and developing countries need guidance on the choice of methods for its measurement. In heavily industrialized areas, such measurements may be needed as part of pollution-abatement programmes, and in countries now undergoing rapid technological development they may be necessary for the making of proper decisions on the site and design of power plants and industries and on the fuel to be used.

The present work discusses some of the fundamental principles on which air sampling and analysis are based and outlines some of the methods now in use in different countries for identifying and estimating the most common pollutants. It was originally submitted, as a draft report on Methods of Measuring Air Pollutants, for discussion by a WHO Scientific Group on the Identification and Measurement of Air Pollutants that met in Geneva on 16-22 November 1965, and was extensively revised in the light of suggestions made by the Group. The revised document was subsequently circulated to a number of experts of international repute in different parts of the world, and the present publication incorporates many of their suggestions. A list of these reviewers will be found on page 7. To them, and particularly to the author, Professor M. Katz, the World Health Organization extends its sincere thanks.

CHAPTER 1

Air Pollution Studies: General Considerations

In selecting methods for detecting and measuring air pollution, consideration must be given to several different factors. One of the most important of these is the objective of the study. Air pollution investigations may be undertaken for any of a number of reasons, some of which are discussed in the following sections.

Investigation of Complaints

Bad odour

Accurate estimation of the concentration of substances responsible for unpleasant odours—which is, in any case, difficult—is often unnecessary, since the source of the odour may be well known or may be clearly indicated by the distribution of the odour. Hydrogen sulfide and thiols (mercaptans) are often responsible for such odours, but many other gases and vapours can be offensive, even in low concentrations. In selecting methods for estimating odorous substances, it should be remembered that if thiols are present, sulfur dioxide may also be present.

Soiling of property

Soiling is caused largely by soot, dust, and other particulate matter. However, particles that are sufficiently heavy to settle quickly are less important—from the points of view of both soiling and health—than those that remain in suspension for long periods. For this reason, the collection of settled dust and fallout in jars is usually less satisfactory than the sampling of suspended matter. Suspended matter may be sampled by high-volume methods or by a paper-tape sampler. The former procedure is particularly valuable, since it lends itself to different degrees of analytical complexity and gives samples that can be stored for subsequent analysis or sent to a reference laboratory. Paper-tape methods require calibration and give only relative results.

Irritation of eyes and respiratory tract

Sulfur dioxide is one of the principal contaminants that may cause irritation of the eyes and respiratory tract, particularly in areas where substantial use is made of coal or fuel oil. If irritation occurs on warm sunny days in areas where motor vehicle traffic is heavy, consideration should be given to the possibility that ozone and other oxidants may be responsible.

Impaired visibility

When impairment of visibility occurs, the concentration of particles and smoke in the air should be investigated, with particular attention to soot if the phenomenon occurs during cold, foggy weather. If it occurs during dry weather, dust or photochemical aerosols may be responsible. High-volume or paper-tape methods should help to identify dusts and measurements of oxidants may be useful for identifying photochemical pollution. However, visibility may be impaired by photochemical aerosols with only small amounts of oxidants present.

Relating Pollution to Sources of Emission

In many communities, sources of pollutants include industrial plants, motor vehicles and trains, home and business heating plants, and waste disposal plants.

Power plants and factories are usually point sources with a fluctuating but definable pattern of emission. Each industry may emit different pollutants, but some form of particulate pollutant is likely to be emitted from all heavy industries. Power plants burning fossil fuels are likely to emit sulfur oxides and related substances, as well as some particulate pollutants and oxides of nitrogen. Sampling of effluents from smokestacks may be of assistance in the preliminary identification of air pollutants. Automatic instruments for recording sulfur dioxide levels are useful in such situations, but a number of samples can often be obtained by non-automatic methods at lower cost. Methods for the determination of oxides of nitrogen, based on the use of the Saltzman reagent, may be useful.

Waste disposal may produce an incredibly complex range of pollutants. However, odour, the density of smoke plumes, and particle sampling are often adequate guides.

Motor vehicular exhaust is likely to be a major source of pollution throughout urban areas. It includes unburned and reformed hydrocarbons, oxides of nitrogen, carbon monoxide, minute amounts of polynuclear hydrocarbons, residues of fuel additives (e.g., tetraethyllead), and other substances. Since carbon monoxide is the principal constituent of motor vehicle exhaust, it is a useful indicator of this type of pollution. Methods for the determination of carbon monoxide include the use of palladium indicator tubes; the costly, but rapid and sensitive, infrared absorption technique; and measurement of the heat of combustion during catalytic oxidation. Photochemical reactions can occur between nitrogen oxides and hydrocarbons from motor vehicle exhaust, leading to the production of photochemical aerosols and ozone. The measurement of oxidants is the preferred method for detecting the products of such reactions.

Household heating systems also may produce an area source of pollution; the nature of such pollution depends on the fuel that is used, and its extent depends on the weather. Sulfur oxides and particulate pollutants are produced by the burning of coal and, to a lesser extent, oil. Such heating systems also emit oxides of nitrogen and particulate soot, the amount of which depends on the efficiency of combustion of the system. The disposal of wastes by means of household incinerators also creates area pollution, but the pollutants are likely to be more numerous and complex than those from heating systems.

Evaluation of Control Programmes

Pollution control may well be the initial effort in a community having an obvious problem—e.g., a community where there is only one industry. However, measurements are of great importance in estimating the effectiveness of such a control programme. For example, if pollution from aluminium-reduction or phosphate-fertilizer works is to be controlled, the concentration of fluorides should be measured at appropriate times and locations (the latter to include areas where sensitive crops, such as forage crops, are grown). Similarly, in evaluating the control of pollution caused by power plants that burn solid fuel, the concentrations of sulfur dioxide and particulate matter should be measured. Since other sources may also emit such pollutants, it is desirable to make background measurements that are not affected by the source being controlled.

Evaluating programmes for the control of area-wide sources (e.g., household heating systems, motor vehicles, incinerators, and widespread industry) involves more difficult problems, and necessitates the use of special procedures.

Relating Effects to their Causes

Atmospheric measurements are indispensable if observed effects are to be traced to the pollutants that cause them. The task of relating cause and effect may be simple (as with soiling caused by soot) or extremely difficult (for example, no one substance has yet been found to be capable, at the concentration in which it occurs in the atmosphere, of causing eye irritation). Cause-effect relationships that are well known include the darkening of paint by hydrogen sulfide; the damaging of vegetation by sulfur dioxide, oxidants, fluorides, and ethylene; and the effect of carbon monoxide on the oxygen-carrying capacity of the blood.

Although this report is primarily concerned with analysis for specific substances, the use of nonspecific indicators and indices is appropriate when it is not possible to relate effects to single pollutants. To some extent such indices of pollution are already in use for estimating the approximate severity of pollution—e.g., " oxidant ", which is not a clearly defined substance. Similarly, eye irritation panels and vegetation exposure methods provide an index of the severity of photochemical pollution. Interference with visibility is also a nonspecific but highly practical measurement that may even help to identify sources of pollution.

Prediction and Evaluation of Health Hazards

It is usually impossible to predict threats of air pollution to health on the basis of laboratory research and experience with occupational exposure, for the following reasons:

(1) Communities, in contrast to groups of workers, contain many groups that may be unusually sensitive to atmospheric pollutants—e.g., children, infants, elderly people, the chronically ill, and those who are unusually sensitive for other reasons.

(2) Most research and occupational exposures occur at a fixed concentration for a fixed (usually brief) period, whereas air pollution exposures may fluctuate in an extreme manner or be nearly continuous.

(3) Air pollution usually results in simultaneous exposure to several pollutants, which may occur in combinations that are difficult to identify (e.g., soots, sulfur oxides, nitrogen oxides, carbon monoxide, and photochemical reaction products). Such combinations have received little attention in the toxicological literature and have been rarely encountered in occupational experience.

Consideration of the factors noted above leads to the following practical guides for atmospheric measurement when health effects are a major consideration.

(1) Pollutants whose health hazard depends on the accumulated body burden should be determined by integrating methods. For example, for the determination of lead, the sampling period should be about a week,

and for carbon monoxide it should be several hours or more. There is no realistic way of interpreting the health consequences of instantaneous maxima of these substances.

(2) Irritant pollutants, such as sulfur dioxide and ozone, should be determined by continuous methods, since high 1-hour or 30-minute maxima may have considerable significance. However, long-term exposure to such irritants may also have important effects on health.

(3) Special studies of health effects that are undertaken when a population has been defined will necessitate specific methods of estimating the exposure of the population within buildings and in the open air. The length of the sampling period will depend on whether acute or long-term effects are being studied. However, it should be noted that long-term effects could be caused even by short-term exposures if these acted as " triggers ". Long-term effects could result from the storage, in the organs of the body, of particulate pollutants that are inhaled or ingested. Atmospheric concentrations of such substances and the amounts stored in different organs should be estimated, and such work should be given high priority.

The uses of atmospheric sampling described above are of great importance, but they are not adequately applied in most programmes. They can provide information of great value for the establishment of guides for air quality.

Prediction of Future Pollution Levels

Most air pollution measurements undertaken in the past were made in situations that had already reached the undesirable stage. More and more frequently, however, air pollution studies are being undertaken in order to predict the increase in pollution that will result from industrial development in a given area. For this purpose, meteorological data and background information on the levels of air contaminants already present are necessary. The pollutants to be measured, and the methods to be used, will depend on the nature of the planned industry. Excellent studies have been undertaken in Great Britain prior to the construction of large coal-fired power plants (sulfur dioxide was the principal substance measured). In developing areas, it is necessary to plan not merely single installations, but entire industrial complexes, together with cities and utility services. By making measurements of existing pollutants, or by using test sources and measuring the resulting pollutant levels, reasonable predictions are feasible.

When agreement has been reached on air quality guides, it may be expected that the guides will be useful for predicting the likelihood of undesirable pollution from proposed new processes and installations.

Nomenclature and Units

If air pollution data from different countries are to be comparable, it is essential that they be expressed in standard units. Furthermore, the origin, constitution, and properties of pollutants, and the meteorological phenomena associated with their occurrence, should be described in precise, unambiguous terms. For a list of recommended units of measurement, see Annex 1.

CHAPTER 2

Air Sampling

PLANNING AN AIR POLLUTION SURVEY

The different factors involved in the planning, scope, and conduct of community air pollution surveys for both routine and research purposes have been reviewed by Rossano.[199] A technical air pollution study of a community or problem area usually requires considerable numbers of personnel and large amounts of laboratory and field equipment for a year or longer. Rossano has listed a number of such studies whose objectives were to determine the origin, magnitude, and causes of air pollution and its effects on vegetation, public health, and property, and to recommend effective action for control. Although considerable progress has been made in this type of research, there is still insufficient understanding of the processes involved in the formation, accumulation, diffusion, dispersion, and decay of air pollutants; of the photochemical and other reactions that take place in the atmosphere; and of the undesirable effects of pollution on living organisms and on the over-all environment.

The fundamental requirement of a technical air pollution survey is the collection of data that form an accurate record of air quality. Such data must cover the content of particulate and gaseous contaminants and its fluctuations in time (e.g., diurnal, seasonal, and annual cycles and long-term trends) and in different areas (e.g., the influence of topographical and geographical factors on the horizontal and vertical distribution of pollutants). To meet these requirements in a large-scale study, a network of fixed sampling stations is usually necessary within and around the survey area. This network may be supplemented by the use of mobile sampling equipment in one or more vehicles.

The American Society for Testing and Materials has published a recommended procedure for planning the sampling of the atmosphere,[22] which covers the broad concepts involved in the analysis of air for pollutants. Since the analysis of air is influenced by factors that are beyond the control of the investigator, the number and size of samples and the interpretation of data must be based on statistical considerations governed by the theory of probability. Consequently, regular observations should be made of all meteorological and other relevant variables; when a sufficient number of observations has been made, the relationship of the factors observed to atmospheric composition can be determined statistically. Mean, median, and maximum values of given constituents observed at different sampling stations may be examined in relation to other statistics. However, the number of samples that must be obtained in a given survey may also be affected by factors other than statistical considerations.

Duration of Sampling Period

Two types of sampling are used in studies of air pollution: short-period or " spot " sampling, and continuous sampling for the evaluation of peak and average concentrations over definite time intervals. Spot samples are usually collected over periods varying from less than 30 min to several hours for specific, well-defined purposes. Such samples have only limited value, except where traffic or other local sources of pollution cause only minor changes in concentration at particular periods of the day. Spot sampling is also useful for the random checking of pollution at many points. In general, pollution levels fluctuate widely, depending on meteorological conditions, topographical features, and various factors associated with the sources of pollution (e.g., the mass rates of emission of pollutants from smokestacks; the temperature, velocity, and density of stack gases; the height of stacks; the distribution of sources; and the downwind distance from the sources to the points where the measurements are made). Consequently, spot sampling cannot give adequate data on the nature and magnitude of an air pollution problem.

Continuous sampling techniques are usually necessary in systematic studies of the nature and extent of air pollution if the data are to be of value for epidemiological surveys, for evaluating the potential hazard to man, animals, or vegetation, and for control programmes. By means of such techniques, maximum and average concentrations over definite time periods can be read from the instrument or calculated, depending on the duration of the sampling cycle. Peak concentrations over relatively short periods (less than 30 min) within a given cycle may also be determined. However, continuous-sampling instruments may lack specificity, and data obtained from them must be frequently checked by calibration and by other, more specific, analytical methods. Continuous sampling may be carried out by the sequential chemical absorption or filtration of measured air volumes, and subsequent analysis of the pollutants, or by means of properly calibrated automatic instruments.

In the interest of comparability, it is recommended that standard collection periods be used in routine sampling measurements; thus, a short-term sample would be one collected during 30 min and a long-term sample would be one collected over 24 h. Under certain circumstances, these periods may be too short (e.g., for the measurement of pollutants whose concentration is too low for them to be detected by a short-term sample) or too long (e.g., if " instantaneous peaks " are important). Furthermore, the choice of the sampling period may depend upon the nature of the compound under study and its stability to oxidation, light, or other factors (e.g., in the analysis of photochemical reaction products or polycyclic aromatic hydrocarbons).

In the determination of " instantaneous values ", or peak concentrations of a pollutant over short-time intervals of less than 15 min, a knowledge of the time-constant of the sampling or continuous-recording instrument is of the utmost importance. The time constant of any sensing device is equivalent to the time required for a reading to attain 63% of its true value, or $1-(1/e)$. Consequently, to determine the true peak concentration the sampling period must be at least four times the time constant of the instrument employed. A method for determining the time constant of a continuous gas sampling and recording instrument has been described by Katz et al.[212]

Samples should be collected under all environmental conditions that affect pollution levels—e.g., all weather conditions. This may necessitate sampling continuously around the clock. There is sufficient evidence that pollution is influenced by diurnal variations in wind speed, atmospheric stability, solar heating and cooling, cloud cover, and other factors. The distance from a source at which gases and small particles (less than 5 microns in diameter) are brought to ground level depends upon the height of the smokestack, the height to which the smoke plume rises, the wind speed, and the stability of the atmosphere. Biological effects, rates of corrosion, and visibility are influenced greatly by the moisture content of the atmosphere and by the interaction of water vapour with condensation nuclei. Natural air cleaning processes, including precipitation and the passage of fronts, materially affect levels of soluble and insoluble pollutants.

The necessary sample size depends principally upon the limitations of sampling devices and analytical procedures and on whether short-period, average, or peak concentrations are to be evaluated. For example, the minimum sample size is governed by the sensitivity, accuracy, and precision of the analytical methods to be used. Frequent short-period sampling is more advantageous than cumulative sampling, since cumulative data can be obtained by the summation of the results of a series of short-period samples. If, after collection, samples are transported to the laboratory for analysis, they must be handled in such a way that their composition when they reach the laboratory is the same as when they were present in the atmosphere.

Location of Sampling Sites

Sampling sites must be carefully selected so as to be representative of the areas under study. For the study of health effects, the sites must be located so that the collected samples represent air that is actually breathed by the exposed population groups. Since concentrations of pollutants vary with altitude, results obtained on the roofs of tall buildings may differ substantially from conditions at ground or breathing level. The necessary number of sampling stations, and their location, depend on several factors, including the objectives of the programme, the size of the study area, the proximity of sources of pollution, topographical features, and weather. A representative number of sampling stations for a given area may be established by means of a preliminary survey, whose objectives should be (1) to gather information on the nature and magnitude of the emissions from principal sources of pollution, (2) to review the available climatological and meteorological data, and (3) to gather data on the concentration of pollutants in areas of severe and slight pollution.

Air sampling networks of different patterns have been used in different studies. For example, geometric patterns were used in the Detroit-Windsor Air Pollution Study, and in the Louisville,[257] Paris,[159] and Nashville [279] studies. In the Aerometric Survey of Los Angeles [188] the sites were selected on the basis of typical trajectories of surface winds, and did not conform to any specific pattern. In the Nashville study, an extremely large number of stations were arranged symmetrically to represent a triangulation system superimposed upon a circular grid, with the centre of the city as its geometric centre. The stations were located at the apices of equilateral triangles whose sides were about 0.5 mile (0.8 km) in length.

Sampling Methods

A wide variety of instruments and procedures is available for collecting samples of particles, aerosols, gases, and vapours. The instruments and methods that are selected will depend on the quantities of pollutants in the atmosphere and on their physical state and chemical properties. Gases and vapours may be studied by means of large-volume samples containing the material in its original state or by methods involving absorption in solvents or solid media or condensation by freeze-out techniques.[a]

METHODS OF SAMPLING AEROSOLS

Aerosols consist of particles, either liquid or solid, dispersed in a gaseous medium. Liquid particles are spherical in shape but solid particles may be spherical, filamentous, or irregular.

[a] A review of air sampling instruments will be found in ref. 17. For reviews of the literature in this field, see refs. 10, 16, 18, 19, 22, 44, 46, 47, 104, 106, 115, 120, 127, 153, 183, 190, 193, 206, 276, 282.

The size of dust, fume, and aerosol particles varies widely, depending on the way in which they are formed and on their stability in the atmosphere as influenced by gravity settling, coagulation, sedimentation, impaction, Brownian movement, electrical charge, and other phenomena.

Generally, by definition, aerosols include solid and liquid particles whose diameters range from about 100 microns to 0.01 micron or less. Particles whose diameters are larger than 100 microns tend to settle out of the air by gravity fairly rapidly. The size ranges of various particles are listed in Table 1.

TABLE 1. APPROXIMATE SIZE RANGES OF AIRBORNE PARTICLES *

Substance [a]	Diameter, microns	
	Minimum	Maximum
rain drops	500	5 000
human hair	35	200
stoker fly ash	10	800
pollen	10	100
plant spores	10	35
red blood cells (adults)	7.5	
cement dust	3	100
water vapour (natural mist, 60-500 microns; natural fog and clouds, 2-60 microns)	2	500
foundry dusts	1	1 000
pulverized-coal fly ash	1	50
metallurgical dust	0.5	100
ground talc	0.5	50
insecticide dusts	0.5	10
bacteria	0.3	35
sulfur trioxide mist	0.3	3
pigment, paint	0.1	5
alkali fume	0.1	5
ammonium chloride fume	0.1	3
oil smoke	0.1	1
salt nuclei (from sea water)	0.03	0.5
colloidal silica	0.02	0.05
metallurgical fume	0.01	2.2
resin smoke	0.01	1
tobacco smoke	0.01	1
normal impurities in quiet air	0.01	1
zinc oxide fume	0.01	0.5
magnesium oxide fume	0.01	0.5
carbon black	0.01	0.3
combustion nuclei	0.01	0.1
viruses and proteins	0.003	0.05
gas molecules	0.0001	0.0006

* Data from ASTM D1357-57 (ref. 22).
[a] Arranged in order of decreasing minimum diameter.

The methods used to collect samples of aerosols for enumeration, measurement, and identification of particles depend to a great extent on the physical properties of the aerosols under study.[a] In air sampling, the

[a] For reviews of the physics of aerosols, see refs. 93, 125, 165, 194, 227.

method is governed largely by the particle size distribution that is of interest.[a]

Two general processes are involved in the formation of aerosols: condensation and dispersion. In condensation processes, clusters of molecules of a substance unite to form particles of colloidal dimensions. The formation of a cloud of liquid droplets when warm air is cooled as it rises into the upper atmosphere is an example of a condensation process. When volatile substances and solids are heated to a temperature at which they evolve vapours and fumes, small liquid and solid particles may be formed as a result of the condensation that occurs on cooling. In the atmosphere, the presence of nuclei and nucleating agents (e.g., ions, polar molecules, fine dust particles, and molecular aggregates of the vapour itself) provide the surfaces upon which vapour can condense to produce an aerosol.

In dispersion processes, relatively coarse matter is split up into fine particles by mechanical grinding, abrasion, and many other processes. Examples of dispersions are the discharge of fly ash from pulverized-fuel furnaces, the emission of dust from foundry shake-outs, the discharge of solids from blast furnaces, the production of dust in the manufacture of cement, and the discharge of particulate contaminants by motor vehicles.

Aerosols and airborne particulate matter may be sampled by any of the methods discussed in the following sections.[b]

Sedimentation

Sedimentation is one of the simplest techniques for the collection of particulate matter, but it is limited in general to particles whose diameter exceeds about 10 microns. It is widely used in the determination of particle fallout or dustfall in community atmospheres. The collectors are usually glass jars or metal or plastic cylinders whose height is about 2–3 times their diameter. After an exposure period, usually a month, the soluble and insoluble material is determined and the total solids collected are expressed in terms of weight per unit area per 30 days. The results are affected by the shape and dimensions of the container and by the presence or absence of aerodynamic shields to deflect wind currents that tend to blow across its open end. Some types of collector, such as the British Standard Deposit Gauge, consist of a funnel or bowl to collect dust and a bottle into which rain can drain for the determination of soluble matter. The data obtained

---

[a] In the study of effects of aerosols on health, particle size also has an important bearing on the penetration and retention of particles in the lungs. Coarse particles are effectively excluded by the nasal and oral passages and only those whose diameter is smaller than 5–7 microns enter the lungs. Particles having diameters of about 5–0.25 microns are effectively retained in the lungs. Particles less than 0.1 micron in diameter are retained in the lungs only to a slight extent, although even these fine particles may be absorbed by the finest branches of the respiratory system. According to Drinker & Hatch,[71] the optimum particle diameter for alveolar deposition is about 1 micron for inhaled dust such as silica, although the optimum size would be smaller for dust of higher density (e.g., about 0.5 micron for lead dust).

[b] For discussions of methods of sampling and estimation, see refs. 10, 84, 104, 127, 193, 196.

by sedimentation methods are not relevant to the elucidation of health problems, and the methods are mentioned here only for review. The limitations of this type of sampling have been described in a recent experimental and statistical study of various types of dust-fall collector.[210] It is obvious that results of particle fall for different areas cannot be compared unless the dimensions and design of the collectors, their location, and other details are rigidly standardized. Procedures for the measurement of dustfall have been formulated by the American Society for Testing and Materials,[22] the Air Pollution Control Association,[2] and the British Standards Institution.[39]

Glass microscope slides or greased plates may also be used in the examination of particles by sedimentation. A method that is widely used for the sampling of pollen suspended in the atmosphere involves collection on a 3 $\times$ 10-cm glass plate or slide coated with petroleum jelly and placed horizontally in a shelter constructed from two stainless steel discs, about 20 cm in diameter, arranged horizontally and spaced about 10 cm apart by three separators. The glass slide is inserted in a holder about 3 cm above the lower disc. The shelter is carried on a rod, about 80 cm in height, with a supporting base, and is usually placed on top of a tall building. After a 24-h exposure period, the slides are stained and the pollen grains identified and counted under a microscope, the concentrations being reported as the number of grains per cm^2 per 24 h. The results are only roughly quantitative.

Settling chambers have also been used to collect particles by sedimentation. Microscope slides are taped to all interior surfaces of a varnished wooden box, one entire side of which is occupied by a door. With the door open, the sample is obtained by passing the chamber back and forth through the air or by fanning the air gently into the box. The particles are allowed to settle on the slides over a period of 16–24 h. The collection of particles is governed not only by gravity settling but also by diffusion of the smaller particles, particularly those of submicron size, to the walls and "ceiling" of the box at a rate that exceeds their settling velocity. Convection currents, set up by even small temperature changes, may bring particles to the walls, on which they may impinge through the action of inertial forces.

Particles collected by sedimentation methods are usually more or less agglomerated, and consequently may not be representative of the original form and size of the particulate matter suspended in a contaminated atmosphere. Equations for calculating the settling rates of particles of different sizes may be found in standard textbooks.

Filtration

Filtration methods and instruments are being utilized on an increasing scale for the collection and evaluation of atmospheric aerosol particles. Filters are widely used in the study of the mass concentration number, optical properties, size distribution, and chemical composition of particles suspended in the air. The choice of filter media and sampling instruments is governed largely by the types of test to be undertaken and the information that is sought.[a] For example, in the determination of mass concentration per unit volume of air, the primary objective is to collect a weighable quantity of aerosol within a reasonable sampling period by means of a high-efficiency filter that has comparatively low resistance to high flow rates. If chemical or spectographic analysis of the material is a primary requirement, the most suitable filter is one that is substantially free from organic or inorganic impurities that might cause interference. Determination of the size distribution and shape of particles requires a membrane filter of such a type that particles will remain on its surface and not penetrate into the interior of the filter bed. For examination in the electron microscope, the membrane should be readily removable by gentle solution in the vapour of alcohol, acetone, or other low-boiling-point solvents without disturbing subsequent deposition of the particles on the Formvar screen of the electron microscope. Evaluation of optical absorbance by the measurement of light transmittance or reflectance requires the use of either paper or plastic membrane filters of uniform thickness and high transmissivity or reflectivity.

Filtration efficiency

Filters may be classified in several groups according to their structure and composition. The efficiency of the different types is discussed in the following sections.

Filter paper

The efficiency of filter paper for aerosol collection is usually a direct function of pressure drop and gas velocity. Its efficiency in collecting particles of submicron size particles may be evaluated by the degree to which it is penetrated by dioctyl phthalate smoke particles, which can be generated with a fairly uniform diameter of 0.3 micron. Filter media of different types vary widely in collection efficiency and resistance to air flow (see Table 2). The efficiency of the less effective papers may be considerably improved by an increase in flow rate, if the resultant increase in pressure drop is not a disadvantage. The efficiency also increases with the loading.

Inorganic-fibre and mixed-fibre filters

A number of highly efficient inorganic filters in the form of mats, pads, and felts have been developed, as well as others composed of mixed glass-asbestos or vegetable-asbestos fibres, primarily for air-cleaning purposes.

[a] For reviews of filter media for air sampling, see refs. 49, 127.

TABLE 2. EXAMPLES OF HIGH-EFFICIENCY FILTERS

Filter	Composition	Efficiency [a] (%)
W. & R. Balston, England		
Whatman GF/C	glass fibre	>99.9
Whatman no. 32	cellulose	99.5
Whatman no. 42	cellulose	99.2
Whatman no. 44	cellulose	98.6
Whatman no. 50	cellulose	97.0
Gelman Instrument Co., USA		
type VM-4	polyvinyl chloride	~100
Acropor, type AN-3000	acrylonitrile/polyvinyl-chloride copolymer reinforced with nylon	~100
type A	glass fibre, no organic binder	>99.99
type E	glass fibre with acrylic binder	>99.99
Hollingsworth & Vose, USA		
HV-70, 18-mil [b]	cellulose & asbestos	99.3
Millipore Filter Corp., USA		
types HA & AA	cellulose esters	>99.9
Mine Safety Appliance Co., USA		
type 1106-B	glass fibre	99.99
type 1106-B	flash-fired glass fibre	99.97
Schleicher & Schüll, Federal Republic of Germany		
Ultrafilter	cellulose ester	99.99 [c]
Schneider-Poelman, France		
type " rose "	esparto & asbestos fibres	99.98 [d]
type " jaune "	esparto & asbestos fibres	97.0 [d]
USSR		
FPP-15-4.5		99.995 [e]
FPP-15-3.0		99.99 [e]
BF		99.9 [f]
AFA RMP		99.5 [e]
AFA RMA-20		97.0 [e]
US Atomic Energy Commission		
AEC-1	asbestos & vegetable fibres	>99.9
CWS-6	asbestos & cellulose fibres	>99.9

[a] It should be noted that, owing to differences in methods of measurement, not all the figures for efficiency are comparable. Unless otherwise noted, the figures given are based on retention of di-(2-ethylhexyl)phthalate particles 0.3 μ in diameter, at a flow rate of 0.1 m/min.

[b] The figure given is for pore size. 1 mil = 1 milli-inch = 0.00254 cm.

[c] Efficiency based on retention of di-(2-ethylhexyl)phthalate particles not exceeding 1 μ in diameter at a flow rate of 0.1 m/min.

[d] Efficiency based on retention of indigo aerosol particles 0.3 μ in diameter.

[e] Efficiency based on retention of decay particles of radon.

[f] Efficiency based on retention of oil aerosol droplets, diameter not stated.

However, the thickness and complex composition of such filters usually render them unsuitable for use in atmospheric sampling. In its study of particulate pollutants in the atmosphere of cities in the US National Air Sampling Network, the US Public Health Service is making extensive use of flash-fired glass-fibre web as a collection medium. When used with a high-volume air sampler, this filter can quantitatively recover the particulate matter from 2000–2500 m^3 of air in continuous sampling over a 24-h. period.

Granular filters

Granular filters may consist of a wide variety of materials, such as porous ceramic tile, fritted glass, and soluble crystalline substances, including sugar and salicylic acid. Although their collection efficiency may cover a wide range of particle size, granular filters usually cannot be used for the collection of aerosol particles whose diameter is less than about 1 micron.

Controlled-pore-size membrane filters

The introduction of molecular sieve membranes of controlled pore size,[a] originally developed in the Federal Republic of Germany as a cellulose ester xergel,[b] has been one of the most important developments in filtration. A number of similar filters, and others composed of polymers, are manufactured in the USA by the Millipore Filter Corporation.[c] The structure, properties, and application of these membrane filters have been described by Goetz.[84] They consist of an effective filtration layer about 0.5 micron in thickness, supported by a dry, spongy gel having low resistance to air flow. Filtration is achieved by deposition of aerosol particles almost entirely on the surface of the membrane. The filters are ashless and completely soluble in alcohols, ethers, and ketones. Their collection efficiency is remarkably high, being greater than 99.9% for dioctyl phthalate (DOP) smoke particles having a diameter greater than 0.2 micron. Types HA and AA Millipore filters show pressure drops of 79 and 62 cm of water at a linear sampling rate of 8.55 m/min.

A wide variety of Gelman [d] filters are now available. Membrane filters composed of porous cellulose esters may be obtained in a wide variety of pore sizes, ranging from 0.004 to 10 microns. The filtration efficiency of type VM-4, whose pore size is 0.80 micron, is about 99.3% for particles having a diameter of 0.05 micron, rising to about 100% for

---

[a] Available from: Schleicher & Schüll, Dassel, Federal Republic of Germany; and from Sartorius-Membranfilter GmbH, Göttingen, Federal Republic of Germany.

[b] " Xergel " is a term that is applied to a membrane filter of uniform composition and controlled pore size that is used to separate colloidal particles from liquids and gases.

[c] Available from: Millipore Filter Corp., Bedford, Mass., USA.

[d] Available from: Gelman Instrument Co., Ann Arbor, Mich., USA.

particles whose diameter is 0.08 micron or greater. The efficiency of Gelman types A and E glass-fibre filters is 99.97% for DOP particles 0.3 micron in diameter. Type G glass-fibre filter shows a low pressure drop and an efficiency of 98.5% in the collection of DOP aerosol. The type A filter is completely free of organic material and may be used for high-volume sampling studies. A mat-like filter made from polystyrene filaments 0.3 micron in diameter combines the advantages of high efficiency with relatively low pressure drop, even under severe loading conditions. At a face velocity of 100 cm/sec, its efficiency in collecting particles 0.3 micron in diameter is about 99.7%.

The physical and chemical properties of such membrane filters render them ideal for collecting particles for microscopic measurement of number and size distribution, for electron microscopic studies of shape and optical properties, and for chemical analysis.

Impingement Methods

Impingement apparatus for the collection of aerosols may be either wet or dry. Wet impingers collect particles by causing them to impinge on a surface submerged in a liquid. Dry impingers, usually referred to as impactors, collect particles by impaction on a dry surface. In both types of apparatus, collection occurs as a result of inertial force, as the particles tend to resist a change in direction when the air stream is deflected by a surface or other obstacle. The efficiency of collection by impingement or impaction is high for particles whose diameter is 1 micron or greater. For collection of submicron particles, the impinger should be operated at high velocities.

The well-known Greenburg-Smith standard and midget impingers are examples of wet collectors. In the former type, the air sample is drawn through an orifice 2.3 mm in diameter and impinges against a flat surface 5 mm distant, with a jet velocity of about 113 m/s at a sampling rate of 28.3 litres/min. In the midget type, air is drawn at a rate of 2.8 litres/min through an orifice 1 mm in diameter and impinges against a flat surface 5 mm distant, at a jet velocity of about 60 m/s. The collection efficiency is low for particles less than 0.7 micron in diameter.

Dry cascade impactors are also available commercially. One type consists of a series of four jets of progressively smaller size, arranged in sequence so that the direction of air flow changes 90° after passage through each stage. Particles are collected on microscope slides placed in front of the jet of each stage.

One of the principal objections to the use of impingement devices for collection of aerosols is that large particles may tend to fracture or separate into smaller particles at the high impact velocities attained, leading to the formation of many more fine particles than were present in the original sample. A theoretical and experimental study of the collection of aerosols by impaction has been described by Ranz & Wong,[184] and both impingement and impaction investigations have been reviewed by Kusnetz.[139]

Electrostatic precipitation

The theory of electrostatic precipitation in the separation of airborne particles has been reviewed by Davies.[69] The particles in the air stream pass between electrodes that are charged at a potential of 12 000–30 000 vDC, and become charged by collision with ions in the electric field. Under the influence of the electric field, the charged particles are transported to a collecting electrode, where their charge is neutralized. Collection of particles at nearly 100% efficiency occurs readily over a reasonable length of gas path with particles ranging in diameter from 10 microns to about 0.2 micron. The efficiency of collection is a function of many factors, including the design of the electrodes and the spacing between them, the velocity and the contact time, the applied voltage, and the electrical conductance of the particles.

Since the effectiveness of a precipitator depends upon the extent to which a charge can be acquired by the particles, it will not remove suspended solids or liquids that can volatilize at its operating temperature. Furthermore, if the applied voltage approaches the breakdown voltage for the gas system, arcing or sparking may occur; consequently, precipitators cannot be used in explosive atmospheres. However, electrostatic precipitation uses a high air flow rate compared with that of many other types of sampling equipment; consequently, it has the advantage of making it possible to collect weighable quantities of particles in a comparatively short time. Electrostatic collectors and their use in particle size analysis are described in ref. 141.

The most important mechanism in the charging of particles whose diameter is greater than about 2 microns is bombardment by ions. For particles whose diameter is less than about 0.2 micron, the controlling mechanism is charging by ion diffusion or the motion of ions induced by the thermal motion of surrounding gas molecules. Both mechanisms are effective in the collection of particles whose diameter is 0.2–2.0 microns. Relatively high collection efficiencies (greater than 98%) are possible for many types of dust and fumes of submicron size. An equation for determining the collection efficiency of an electrostatic precipitator is given in Annex 2.

Thermal precipitation

If a gas containing particles in suspension is brought into contact with a heated surface, a dust-free space develops between the surface and the particles, owing to the action of a thermal force that causes the particles to migrate from a zone of high temperature to one of lower temperature. Both theoretical [78] and experimental [198,220] studies have provided an explanation of this phenomenon. In a thermal gradient, fluid molecules tend to flow around suspended particles from the colder to the warmer region; reaction to this flow results in the movement of the particles towards the cooler region. A high temperature gradient, greater than about 750° C/cm, is essential for effective precipitation and many commercial instruments operate with a thermal gradient greater than 3000° C/cm. According to the theory of Epstein,[78] the thermal force is directly proportional to the gas temperature gradient and to the particle diameter.

For atmospheric sampling, the collection efficiency of thermal precipitators is almost complete over a wide range of particle sizes from less than 0.01 micron to about 10 microns. Collection by this means is ideal for the direct determination of the number and size distribution of aerosol particles by microscopic examination of the material deposited on slides. The particles are deposited gently in their natural state without shattering. Several models of thermal precipitator are available that contain a hot wire, which is heated electrically so as to establish a uniform temperature gradient, suspended near a glass microscope slide. Suction is provided by water displacement or by a small vacuum pump. Another type consists of two parallel discs, one heated and the other water-cooled; the air enters at the centre of the heated disc and flows radially to the edges. The unit is self-contained, with a cooling pump, air pump, and flow meter, and is designed to operate at the comparatively high flow rate of 500 cm³/min.[135] However, most thermal precipitators operate at flow rates of 7–20 cm³/min.

Centrifugal methods

Most centrifugal sampling devices are constructed on the principle of the cyclone. The dust-laden gas, moving at high velocity, is directed tangentially into a cylindrical chamber, in which it is forms a confined vortex. Centrifugal force tends to drive the suspended particles to the wall of the cyclone body, from which they drop into a dust collection chamber. An axial outlet is provided for the cleaned gas. Various modifications of this principle may be used—e.g., *(a)* a tangential inlet with peripheral dust discharge; *(b)* an axial inlet with axial dust discharge; *(c)* an axial inlet with peripheral dust discharge.

Cyclonic collectors are not, in general, highly efficient except for particles greater than 5 microns in diameter. For smaller particles, the collection efficiency is usually about 50–80% by weight. The efficiency increases with increasing inlet velocity, but at a lower rate than the resultant increase in pressure drop. Practical considerations set a limiting value for the inlet gas velocity above which the ensuing turbulence may cause a decrease in collection efficiency.

Goetz and his co-workers [85,86] have developed an aerosol spectrometer for the quantitative separation of submicron particles into different size ranges. The instrument, which consists of a rapidly rotating helical channel inside a removable conical cover, is a special type of conical centrifuge. The rotation of the helix imparts a high centrifugal acceleration (about 20 000 g) to the particles, which are separated in accordance with their density and diameter according to Stokes' law, and which collect on foil fixed to the inside of the cone. By calibration with monodisperse aerosols of known particle size and mass, a quantitative relation has been established between the size of particles and the length of foil on which they are deposited. Another function relates the size of particles to the deposit area occupied by particles of equal size. The deposit may be divided into fractions of different maximum particle size by cutting the foil into a series of annular sections. A specific mass fraction in each of these sections, or the total mass, may subsequently be determined by photometric or chemical analysis. The device has been used by Goetz in photochemical studies of aerosol particles as small as 0.2 micron in diameter.

A conical centrifuge has been developed by Sawyer & Walton [218] to separate and deposit airborne particles according to their size, depending on their relative settling velocities. The instrument consists essentially of two concentric cones mounted on the same vertical axis. Both cones have an apical angle of 60° and are separated by a distance of 0.58 cm. The inner cone, which is solid, is rotated at a speed of 3000 rev/min. An air sampling jet, through which air passes at a rate of 25 cm^3/min, is attached to a fixed cylindrical housing, and rotation of the cones causes the air sample to circulate through the annular space between them. Aerosol particles are deposited on the outer cone, the larger particles being deposited near the apex and the smaller ones towards the base. The deposited particles are collected for microscopic examination on glass slides inserted in two diametrically opposite slots in the outer cone.

METHODS OF SAMPLING GASES AND VAPOURS

Methods for the sampling of gases and vapours may be summarized as follows:[a]

(1) Sampling of atmospheres in their original state, substantially without concentration of the contaminants.

(2) Sampling of atmospheres with concentration of gases and vapours by the following methods:

(*a*) absorption in chemical reagents or solvents contained in various types of absorption apparatus;

---

[a] Recommended methods for such sampling are discussed in ASTM D1605–60 (ref. 22).

(*b*) adsorption by activated carbon, silica gel, or other solid adsorbents;
(*c*) condensation at low temperatures by freeze-out techniques; and
(*d*) adsorption at low temperatures on a gas-liquid partition substrate.

Sampling of Gases in their Natural State

The apparatus required for sampling gases and vapours in their natural state (i.e., without further concentration) is relatively simple. In special circumstances, particularly for the analysis of gases that occur in the atmosphere in relatively high concentrations (e.g., carbon dioxide), the sample may be collected in a bulb of known volume that has previously been evacuated to a pressure of a few mm Hg. If such a low pressure cannot be obtained, the glass bulb, prepared in an uncontaminated atmosphere, may be evacuated to a known pressure as measured with a mercury manometer. The sample volume may be calculated from the pressure in the evacuated bulb, the barometric pressure, the temperature, and the volume of the bulb. After collection of the gas sample, the bulb may be sealed by forcing a wax-filled cartridge over its broken tip. Samples may also be collected directly in a measured volume of liquid reagent in an evacuated glass bulb.

Samples of relatively inert gases may be obtained by filling containers by gas or liquid displacement. To fill a container by gas displacement, it is essential that the air in the container be completely replaced with the gas sample. This is accomplished with a double-acting rubber-bulb aspirator or a foot pump. Large bags of nonpermeable polyethylene, polyvinyl, or Mylar film (often aluminized) have occasionally been used for air sampling. The collapsed bags are first rolled up tightly to exclude air and are then filled by means of a small electric blower. Alternatively, the collapsed plastic bag may be inserted into an air-tight fibre container with the mouth of the bag open to the atmosphere; the gas sample is collected by slow evacuation of the fibre container at a controlled rate. Samples of up to 25 litres may be collected in this manner. The technique is suitable for use with gases and vapours that neither react with, nor diffuse through, the material of which the bag is made.

Absorption Methods

An assembly for sampling gases by absorption usually consists of the following parts, arranged in the order given: a short probe or sampling line of glass, stainless steel, or nonreactive plastic tubing; a filter for removal of particulate matter; a simple gas bubbler, an impinger, or some other type of absorption apparatus; a gas meter; a pressure gauge or manometer and a thermometer; appropriate valves; and a pump. Some of these parts, such as the filter, gas absorber, meter, and manometer, must be thoroughly tested for efficiency or accuracy. For maximum accuracy, the temperature and pressure conditions under which the sampling is conducted must be measured.

Gas metering devices may measure either volumes or flow rates. Volume meters may be of the wet type or the dry diaphragm type. Although wet meters are more cumbersome than dry meters, they are more precise than the latter. Rotameters, capillary or orifice flow meters, pitot tubes, and nozzles measure flow rates but they must be calibrated (e.g., with a standard gasometer). Such rate meters require close observation, accurate measurement of the sampling time with a stop-watch, and uniform operation of the pump to prevent excessive variations in flow. Volume meters, on the other hand, are not subject to errors if the sampling rate varies. Orifices, pitot tubes, and nozzles are equipped with means for differential pressure measurement.

As the sampling assembly introduces resistance to gas flow, with a resultant pressure drop, it is advisable to install a pressure gauge or manometer after the volume or rate meter in order to provide data for volume correction. If the flow rate is reproducible, as when the resistance of the sampling assembly remains constant, it may be calibrated against the pressure drop on the upstream side of the pump. It is advantageous to use a constant-flow pump.

A constant rate of flow through the assembly can be achieved by means of a calibrated orifice and controlled absolute pressure. Rate of flow becomes constant and is proportional to the upstream pressure when the downstream pressure is equal to, or less than, 53% of the upstream pressure. Consequently, to obtain a constant flow through the sampling assembly, the orifice and pump must be capable of maintaining a downstream pressure well below 50% of that of the atmosphere. However, for dependable results, the resistance of the assembly to gas flow must be reproducible.

Many types of mechanical gas pump (e.g., rotary, diaphragm, reciprocating, cycloid, and centrifugal) are available in a range of sizes. Such pumps are usually equipped with electric motors, of which the induction motor is preferred because of its capacity for uniform operation under varying load conditions. Where electric power is not available, hand pumps may be used. These may be large glass syringes, double-acting rubber-bulb aspirators, or piston pumps. Such hand pumps are not suitable or convenient for sampling large volumes or when constant gas flow must be maintained. Aspirators may be operated either with water or with air under constant pressure. A simple siphon may also be used for small sample volumes or when sampling must be conducted at very low flow rates. The suction tube of the siphon bottle should be adjusted so that the flow of gas remains uniform as the liquid level falls.

Absorption in liquids

The extent to which gas is absorbed in liquids is governed, in general, by the volume of the sample and by the following factors:

(*a*) the degree of contact between the gas and the liquid, as determined by bubble size;

(*b*) the duration of contact between the gas and the liquid, as measured by the rate of gas flow and the length of the gas path through the liquid;

(*c*) the transfer coefficients or rates of diffusion between gas and liquid phases;

(*d*) the degree of solubility of the gas in the liquid;

(*e*) the rate of chemical reaction;

(*f*) the concentration of the reactant in the absorbing solution; and

(*g*) the temperature.

The rate of gas absorption per unit area of liquid interface is directly proportional to the difference in concentration across the path of transfer —i.e., the difference between the concentration in the gas stream over the liquid interface and the concentration in the gas phase next to the liquid surface. In a pure absorption process the amount of gas that can be absorbed by a liquid solvent is limited to that required for thermodynamic equilibrium. If the solubility of a gas in a liquid is directly proportional to the partial pressure of the gas over the liquid, the equilibrium follows Henry's law and yields a straight line relationship. This relation is true at low concentrations over a specific temperature range. With increasing temperature, the solubility of the gas in the solvent is reduced so that a given amount of solvent will absorb less gas.

At equilibrium conditions for diffusion, if the gas in the liquid absorbent forms an ideal solution, the gas absorption curve is represented by the relation $P/C = H$, where H is the Henry law constant and P is the equilibrium partial pressure (in atmospheres) of soluble gas over a solution whose concentration of solute, in moles per litre, is C.

The usefulness of absorption sampling techniques may be greatly increased by the use of chemical reagents (e.g., for oxidation or acid-base neutralization) to reduce the volatility of a contaminant substantially to zero. For acceptable accuracy, the efficiency of absorption should in general be higher than 95%.

A wide variety of absorbers have been used in gas sampling—e.g., simple impingers or bubblers; scrubbers equipped with fritted glass or other types of diffuser; spiral-type absorbers; towers packed with beads or rings; fine-spray scrubbers; and countercurrent columns. The approximate range of usefulness of different absorbers is indicated in Table 3.

TABLE 3. ABSORBERS FOR AIR SAMPLING

Type of absorber	Absorbent capacity (ml)	Sampling rate (litres/min)	Remarks
simple bubbler	10-100	1-30	general purpose; short contact between gas and liquid
standard impinger	50-100	28 or less	useful for readily soluble gases and vapours
midget impinger	10-25	2.8	useful for readily soluble gases and vapours
scrubber with fritted-glass or other diffuser	25-100	0.5-20	good gas-liquid contact, but diffuser has a tendency to plug
spiral scrubber	10-100	0.04-0.5	effective only at low flow rates
packed tower	5-50	0.5-2.0	variable resistance; efficient only at low flow rates
spray absorber	50-100	1-30	capacity of absorbent depends on design and size of absorber; useful for large-volume sampling
countercurrent gas-liquid flow contact	2.5-4[a]	5-20	high gas flow, low liquid flow

[a] ml/min (for SO_2 absorption).

Adsorption

In adsorption sampling, the gas or vapour contaminant is removed from the air sample by contact with a porous solid such as charcoal, activated carbon, activated alumina, silica gel, or an activated earth. From the standpoint of electrical characteristics, solid adsorbents may be classified as either polar or nonpolar. For gas sampling, polar substances have the disadvantage that they adsorb moisture from the atmosphere. Nonpolar substances (e.g., charcoal and activated carbon) can adsorb organic gases or vapours in preference to atmospheric moisture. By the suitable choice of solid adsorbents, it is possible to devise a system for selective adsorption of a mixture of contaminant gases or vapours, and this method is used extensively in column chromatography. Furthermore, each type of adsorbent can be impregnated with chemical compounds for the chemisorption of specific gases or vapours. The collection of such contaminants by purely physical adsorption depends on the temperature and the volume and retentivity of the adsorbent bed. Specifications for activated carbon and its retentivity for a large number of chemical substances in a dry airstream at 20° C and 760 mm are listed in ref. 22.

At ordinary temperatures activated carbon virtually does not adsorb true gases (e.g., hydrogen, nitrogen, oxygen, carbon monoxide, and methane) whose critical temperatures are lower than about 220° K and whose boiling-points are lower than 120° K. Vapours having boiling-points of 173–273° K and critical temperatures of 273–423° K are adsorbed to a moderate degree at ordinary temperatures, but may be concentrated by using a thick carbon bed maintained at a low temperature by refrigeration. Chemical impregnation of the carbon also renders the adsorption of such substances quantitative. Vapours having higher molecular weights and boiling-points above 273° K are readily absorbed and retained by activated carbon at ordinary temperatures.

For subsequent analysis, the adsorbed gas or vapour must be desorbed from the solid adsorbent. Desorption from charcoal or activated carbon may be accomplished by using superheated steam to displace the adsorbed material or by heating the carbon in vacuum and distilling the material into a cooled trap. Vacuum desorption is preferable and is carried out by placing the carbon in a suitable vacuum distillation system that is connected to three receivers in series, which are immersed in baths of ice and salt, dry ice and acetone, and liquid nitrogen, respectively. The carbon is maintained at a temperature of 200–250° C and distillation is carried out at a pressure of about 1 micron Hg until practically no more material can be collected in the receivers. Less drastic methods are used to desorb substances from silica gel or activated alumina. A carrier gas, preferably inert (e.g., nitrogen), may be passed through the adsorbent column at ordinary temperatures or the carrier gas may be heated to temperatures up to 250° C. Extraction with solvents (e.g., water, alcohols, acetone, cyclohexane, and benzene) may also be used, depending on the chemical properties of the substances to be recovered.

Condensation

Condensation or freeze-out methods have the advantage of concentrating contaminant gases or vapours and preserving them in their natural state, without chemical reaction. This method of sampling is particularly useful when the concentration of a given gas in the atmosphere is extremely low and when highly concentrated samples are required for examination by mass spectrometry, chromatography, infrared spectrophotometry, and other physical or chemical methods. In low-temperature sampling, prefiltration is essential to remove suspended particles. The air sample will contain water vapour, hydrocarbons, and other gaseous contaminants that readily condense to the liquid state at the temperature of ice-water or mixtures of ice and salt. Such substances, if allowed to remain in the sample and to pass into the colder traps, would freeze to the solid state and plug the apparatus. For this reason, they should be removed by a glass-wool plug or other filter and added to the condensed sample fractions for subsequent analysis. The choice of refrigerants is usually restricted to those that maintain a constant temperature by a change of phase in the cold bath. In general, the vapour pressure of condensed gases or vapours should be about 1 mm or lower at cold-trap temperatures, to prevent significant evaporation during sampling. Liquid oxygen is usually the coldest refrigerant used for the sampling and condensation of atmospheric gases and vapours. A typical series of refrigerants for sample collection by condensation is as follows: ice and water (0° C), ice and salt (–21° C), dry ice (–79° C), liquid air (–147° C), and liquid oxygen (–183° C) contained in Dewar flasks. Apparatus and procedures for condensation sampling have been described by Shepherd et al.[226] and Barnebey.[25]

Samples collected in the field must be sealed and held at low temperature until required for analysis in the laboratory. The Shepherd trap may be used as a flask for isothermal distillation of a condensed sample at progressively higher temperatures. The distillate is passed through a drying tube containing Ascarite to remove carbon dioxide and then into a condenser immersed in liquid nitrogen. The fraction collected in the condenser may be evaporated directly into the gas inlet system of a mass spectrometer or gas chromatograph.

CHAPTER 3

Photometric Methods for the Determination of Particles

Light Scattering

A number of methods of determining the concentration of particles, based on their absorption or scattering of light, have been devised. Methods of measuring the number, concentration, and size of atmospheric dust particles *in situ* by light scattering techniques are of particular interest. Sinclair [227] has discussed the physics of light-scattering by small suspended particles in relation to their physical properties and the wavelength of incident radiation. The amount of light scattered depends upon the concentration, size, refractive index, shape, and colour of the suspended particles. For most spherical particles and droplets, a diameter of about 0.4–1.2 microns is the size that is most effective (in terms of scattering area per gram) in scattering light and obscuring visibility. Particles of all sizes may reduce visibility by absorbing light, but the very small absorbing particles are most effective in this respect. The haze that is often observed over communities and industrial areas is associated with the presence of particles with diameters near the lower range of the wavelength of visible light. If such a haze consisted of particles of unit density, 0.5 micron in diameter, at a concentration of 0.1 mg/m^3, visibility would be reduced to about 3 miles (5 km). In general, forward scattering of light by submicron particles is greater than backward scattering. When the particle radius is equal to the wavelength of light (about 0.4–0.7 micron), the ratio of forward to backward scattering is about 1:1000 or more.

O'Konski & Doyle [166] have described a light scattering instrument or particle-counter photometer that measures the total light scattered at 90° by individual particles in an air sample. The instrument also counts the number of particles of ten different size ranges and records the data. Although complex, the device has the advantage that suspended dust particles as small as 0.25 micron in diameter can be measured without disturbing the aerosol. By appropriate design of the illumination source, light collection lenses, coaxial flow system, and photometric accessories, a high resolving power is attained for particle-size discrimination. Stray light is virtually eliminated and the background is limited to the light scattered by air molecules. The instrument is calibrated by establishing a curve showing the relation between particle size and pulse amplitude for aerosols of known particle size, using a standard aerosol generator.

An instrument has been designed by Sinclair [228] for the measurement of the mass concentration of aerosol particles and their total forward scattering of light. The device continuously records the mass concentration in the atmosphere of particles about 0.05–40 microns in diameter. Larger particles are excluded by a 325-mesh (44 micron) screen or other filter. The filtered air sample is passed at a rate of about 23 litres/min through a dark-field chamber that is illuminated in such a way that only small-angle, forward scattered light from the aerosol particles strikes a photomultiplier tube. The amplifier current is measured and recorded. The air is then passed through a weighed molecular membrane filter to collect the particles, which are weighed to obtain the mass concentration. Variations in mass concentration during the sampling period may be estimated from the recorded readings by reference to a calibration curve. If calibrated with DOP aerosol particles 0.3 micron in diameter, the instrument has a mass concentration range of about 6–100 000 $\mu g/m^3$.[a]

Transmission and Reflectance

In recent years, wide use has been made of methods of estimating suspended matter and smoke in the atmosphere that are based on measurement of the light transmission or reflectance of deposits on paper or other membrane filters. In England, Owens [224] developed an automatic sampling instrument for the measurement of smoke concentration in the atmosphere as long ago as 1918. In this device, a known volume of air was drawn through a small area on the periphery of a circle of filter-paper, and the shade of the deposit was compared with a set of standard shades numbered from 1 (very light grey) to 20 (very dark grey). The filter-paper was moved to a new position every hour by a clock and weight mechanism, and suction for drawing the air sample was provided by a siphon and a water reservoir holding 2 litres of water. Although the instrument was relatively cumbersome and gave results that were only very roughly quantitative, it was used for many years in London and other cities in Great Britain.

More refined instruments and methods have been described within the last decade by many investigators.[b] Automatic filter-paper samplers are available commercially and combine such features as a motor, a pump, an air-flow meter, a sampling head or clamping device to hold the inlet securely

[a] A proposed method for measuring mass concentration and light scattering, based on Sinclair's procedure, is described in ASTM D1899–61T (ref. 22).

[b] See refs. 96, 99, 103, 115, 128, 134, 147, 161, 185, 211, 214.

against the filter-paper, an interval timer, and a device for setting the sampling cycle and advancing the roll of paper to a new position. The instrument is usually supplied with a meter for measuring the transmission or reflectance of the circular stains.

Katz et al.[128] have described a procedure in which the transmission or reflectance of the stains produced by an automatic paper sampler is measured and in which an air sample is simultaneously drawn through a highly efficient dried and weighed flash-fired fibre-glass filter in a high-volume filtration apparatus for the determination of mass concentration and for subsequent chemical analysis. For the spot samples, air is drawn at a rate of about 7 litres/min through a circle of paper 2.5 cm in diameter. With the high-volume sampler, a fibre-glass filter measuring 20 $\times$ 25 cm is used, the flow rate is 0.7–1.4 m^3/min, and the sampling period is usually 24 h or a sufficient period of time to collect 50–250 mg of particulate matter. A standard test for particulate matter in the atmosphere, based on this procedure, has been adopted by the American Society for Testing and Materials.[22] For transmission or reflectance measurements, monochromatic light with a wavelength of about 400 millimicrons is recommended. The spot sampling probe should be equipped with a funnel covered with a 325-mesh screen (a sieve opening of 44 microns) with a small cyclone dust trap, so that particles larger than about 40 microns in diameter are excluded from the inlet of the automatic filter-paper sampler.

Measurements of the transmission or reflectance of spots on filter-paper can be converted into equivalent mass concentrations of smoke or aerosol particles in the atmosphere only if the instrument is frequently calibrated by means of gravimetric estimations conducted simultaneously on similar air samples. Optical measurements of spots made at different sampling sites may be compared only if the sampling periods are of equal duration and if the atmospheres at the different sites are similar in soiling properties (e.g., the size, refractive index, and colour of the smoke or dust particles). The weight concentration of particles is influenced by their density and varies directly as the cube of their diameter. On the other hand, if, in simultaneous determinations of smoke concentrations in the same air, the sampling time is increased, values for the soiling index or for the absorbance of the spots (in terms of absorbance units per linear units [a] of air sampled) may be decreased. This is a result of the accumulated deposition of particles on the filter-paper spots and of the fact that when the concentration of particles on the paper is high, the absorbance or reflectance is not linearly related to the number or mass concentration of the particles. Even when the concentration is low, the assumption of a linear relation gives only a first-order approximation. For a discussion of absorbance and reflectance, and their relationship to the number, size, and mass of particles, see Annex 3.

[a] Linear units of air sampled = volume of air per unit area of filter-paper.

Rondia [196] conducted a study of the optical density of smoke and dust particles recovered by filtration (through Whatman GF/A glass-fibre and other filter media) from the atmosphere of Liège, Belgium, to evaluate the suitability of transmission, reflectance, and fluorescence for the estimation of concentration. A statistical analysis of the results showed them to be influenced by numerous factors, such as the nature of the filter media, the wind direction and other weather conditions at the sampling sites, the composition of the smoke particles, and seasonal and other variations in industrial and domestic activities. Rondia concluded that the reflectometric method shows less variability than the other methods, but recommended that the quantitative estimation of suspended particles be supplemented by a qualitative study of the nature of the filtered products.

Somewhat similar long-term studies have been conducted by Pelletier [33,176] in France and by Craxford et al.[63] in England on the accuracy of smoke sampling techniques and the significance of the results. These investigators have also adopted the reflectance method in preference to other methods.

In 1964, a working part of the Organization for Economic Co-operation and Development concluded that a procedure of smoke-stain estimation, comprising filtration of the suspended particles and measurement of the darkness of the resulting stain by a reflectometer, can give a valid assessment of the dark material in the smoke in terms of the concentration of a "standard" smoke.[153] The report recommended that a standard procedure be adopted for such estimation, and included a proposed standard procedure and a series of proposed calibration curves. It also made the following recommendations:

> A simple procedure for collection and measurement of the suspended material in the air is by the use of a filter and estimation of the darkness of the resulting smoke stain with a reflectometer.
>
> For all general types of smokes with an intrinsic darkness index within the range 72 to 96 (28 to 4 units of reflectance, Whatman No. 1 paper, Eel reflectometer [adjusted to read 100 on a single sheet of clean Whatman No. 1 filter-paper placed on a standard white ceramic tile]), an established form or shape of smoke calibration curve, relating darkness of stain with surface concentration, should be universally adopted.
>
> If smoke concentrations are required in absolute gravimetric units, the scale of surface concentration to be attached to the ordinate of this curve will differ from place to place and time to time, and can only be determined by separate gravimetric measurements. Unless concentrations in absolute units are specifically required, it is recommended that an agreed scale of surface concentration be adopted and measurements expressed in units of equivalent standard smoke (e.s.s.).
>
> The above units are defined by an agreed scale of surface concentration such that a concentration of 20 µg/cm^e of equivalent standard smoke produces a darkness index 25 when measured with an Eel reflectometer on a [2.54-cm-diameter] Whatman No. 1 filter-paper.[a]

[a] Ref. 153, p. 18.

Experimental work on "cumulative" stains leading to the above recommendations was carried out in France, the Netherlands, and the United Kingdom. Similar work from other sources was also noted, including the studies of Katz et al. and Rondia referred to above.

> The experimental data suggest that for a wide variety of smokes, a single curve is sufficient to correlate darkness index and surface concentration, if the latter is expressed in terms of a standard smoke. The smokes examined varied in intrinsic darkness from 72 to 96 units where the intrinsic darkness is the maximum darkness index that can be obtained, however great the surface concentration of the smoke in question, and is an indication of the characteristic nature of the smoke.[a]

In discussing the relationship between weight and darkness, the report came to the following conclusion:

> The relation between total weight of suspended matter and darkness of stain is very variable and smoke stains cannot be used to measure total weight. The relation between dark material arising from combustion and darkness of stain is less variable and within useful limits the stain may be used as a measure of this material.[b]

---

[a] Ref. 153, p. 17.
[b] *Ibid.*, p. 16.

CHAPTER 4

Gas Analysis

A considerable number of nonautomatic methods for the analysis of gases and vapours that may be found as air pollutants have been developed in different countries. Some of the methods have been subjected to more or less rigorous examination to establish their sensitivity limits, the reproducibility of their results, and their accuracy. Various scientific and technical groups, including international bodies such as IUPAC, OECD, and WHO are interested in standardizing such techniques or in reaching international agreement on recommended practices. Such agreement is vitally important in the determination of substances by methods that are more or less empirical. Since pollutant gases and vapours co-exist in low concentrations in the air, and since the sample must be absorbed in some aqueous reagent, there is a distinct possibility, in most sampling and analytical procedures, of interaction between acids and bases or of the occurrence of oxidation-reduction reactions with the reagent. For example, titrimetric methods may measure only the resultant acid-base balance; colorimetric procedures may be affected by interfering substances that tend to enhance or diminish colour development; or only the algebraic sum of the chemical properties of various absorbed pollutants may be measured.

Methods that are highly specific for given compounds are limited to the measurement of intrinsic properties of the molecule, to the formation of unique reaction products or complexing agents, or to the prior removal of interfering substances by precipitation, reaction, distillation, dialysis, or filtration. When such procedures are not available, the analytical techniques and methods of calibration used must be rigidly standardized if the results obtained in different areas are to be comparable. The over-all nature of community air pollution and of its sources has an important bearing on the validity of nonspecific analytical data.

With respect to their pollution potential, atmospheres may be broadly classified as either reducing or oxidizing. Pollution will be predominantly reducing, with sulfur dioxide as the most important gaseous contaminant, in communities where the predominant domestic and industrial fuels are coal and oil, and where industrial operations (e.g., iron and steel manufacturing, smelting, oil refining, the production of electricity from fossil fuels, and pulp and paper manufacturing) discharge large quantities of oxidizable sulfur compounds into the atmosphere. On the other hand, pollution is likely to be oxidizing, as in Los Angeles, if the fuel requirements of an urban area are met by gaseous or liquid fuels of negligible sulfur content and if the principal pollutants are hydrocarbons and other organic substances, oxides of nitrogen, and carbon monoxide discharged from motor vehicles. Some areas may have both types of pollution, the predominant one depending on the time of year, the weather, or both factors.

Investigations within recent years have demonstrated the complexity of the photochemical reactions that occur in oxidizing atmospheres under the influence of sunlight. The products of such reactions cause eye irritation, reduced visibility, and damage to vegetation, and one of them—ozone—accelerates the deterioration of rubber goods.

A large number of data are available on the concentration of hydrocarbons, aldehydes, oxides of nitrogen, ozone, and other oxidants during days when eye irritation occurs in Los Angeles. The nature of these diurnal variations, observed by many workers, has been elucidated by Leighton.[142] Generally, the first change noted in the early morning in such areas is an increase in nitric oxide and hydrocarbon concentrations, doubtless as a result of the morning increase in motor vehicle traffic. Soon after sunrise, the concentration of nitric oxide begins to decrease and the level of nitrogen dioxide increases. This process continues for several hours, during which the nitric oxide concentration is reduced to a very low level, and at the end of this period the ozone concentration starts to increase. It is believed that the conversion of nitric oxide to nitrogen dioxide and the resultant build-up of ozone are the result of a reaction that takes place between hydrocarbons (chiefly olefins) and nitrogen oxides under the influence of sunlight.

The ozone level usually reaches a peak around noon, or shortly thereafter, and then begins to decrease. This variation in ozone concentration in the atmosphere is brought about by the continued emission of pollutants that react with ozone, by a decrease in solar irradiation, and by the improvement in ventilation that occurs as a result of increasing turbulence during the afternoon. Exceptions to this general phenomenon arise during periods of intense inversion in the Los Angeles basin, when poor natural ventilation and other factors influence the concentration of aldehydes, hydrocarbons, oxides of nitrogen, ozone, and other oxidants. Usually, on days when eye irritation occurs, there is a mid-morning peak in the concentration of hydrocarbons, due to the heavy morning traffic. The level of aldehydes rises gradually throughout the morning, primarily as a result of a photochemical reaction between nitrogen oxides and olefins. During the afternoon, the rapid decrease in ozone concentration is accompanied by a similarly rapid decrease in aldehyde level. It is believed that this decrease in aldehyde concentration is the result of direct photolysis and a combination of other removal processes, such as natural ventilation and reactions with oxygen atoms, free radicals, and ozone produced photochemically.

There is increasing evidence that the oxidizing or photochemical type of air pollution occurs in many other urban areas in different parts of the world. The following substances are capable of absorbing visible and ultraviolet solar radiation in the photochemically active region: oxygen, nitrogen dioxide, sulfur dioxide, ozone, nitric acid, nitrous acid, alkyl and other nitrates, alkyl nitrites and other organic nitrocompounds, aldehydes, ketones, peroxides, acyl nitrites, pernitrites, and particulate matter. Consequently, the possibilities for the formation of diverse, complex reaction products in primary and secondary photochemical processes are extremely great.

From the foregoing discussion it is clear that, in selecting methods for the chemical analysis of gases, careful attention must be given to the composition of pollutants as emitted from the source and to the atmospheric reaction products that occur in the environment. If the large number of other organic and inorganic substances that may be present in the air as dusts, fumes, or particles are also taken into account, the question of specificity and possible interference in analytical methods assumes major importance.

A number of nonautomatic methods of sampling and analysing gases are outlined in Annex 4. All these methods have been used for the study of air pollution problems and for aerometric surveys in different countries, but they have not been adequately evaluated. Intensive collaborative testing of procedures has been undertaken in a number of countries, including the Federal Republic of Germany, France, the United Kingdom, the USA, and the USSR.

In evaluating and selecting methods for the analysis of gases, consideration must be given to the following factors: the efficiency of sample collection; the stability of the reagents and reaction products; the ease of calibration of the procedure; and the degree of simplicity, specificity, sensitivity, reproducibility, and accuracy of the procedure. Factors that affect sensitivity and accuracy (e.g., collection efficiency, stability of reagents and products, speed of reaction, temperature coefficient, and the influence of interfering substances) should be known or capable of control. Serious errors in analysis may result from the use of absorbent reagents that are sensitive to oxidation, light, and temperature or that form unstable reaction products, particularly if there is a considerable delay between sampling in the field and subsequent analysis in the laboratory. The effect of interference may be eliminated by the use of selective or specific reagents or by prior absorption or chemical reaction. If this is not possible, it may be desirable to use a nonspecific method based on the common chemical properties of a group of compounds, if all of them produce a similar biological effect.

As the range of concentration of pollutants in the atmosphere is, in general, extremely low, the sensitivity of the analytical procedure is of major importance in the detection of minimal quantities and in the determination of short-period average and maximum concentrations. To a certain extent, the usefulness of a method of low sensitivity may be increased by sampling a larger volume of air. Both sensitivity and accuracy may be determined by suitable calibration techniques, measured quantities of the pure compound being introduced into the analytical apparatus, at levels corresponding to those existing in the atmosphere, by suitable dilution methods. The diluted mixtures are checked by separate quantitative analyses, and the accuracy of the method is determined by noting the extent of the deviation between the true value and the measured values.

Automatic Sampling and Monitoring Instruments

The use of automatic instruments for sampling or monitoring air pollutants has expanded rapidly within recent years, although Thomas & Abersold [254] devised a continuous monitoring instrument for sulfur dioxide as early as 1929 and several instruments of this type were employed by Katz et al.[129] in the 1930's in studies of the effects of this gas on vegetation in Trail, British Columbia. This trend has resulted largely from the realization that the concentration of pollutants is markedly influenced by meteorological conditions and may show diurnal and seasonal fluctuations, reflecting not only the effect of natural phenomena but also that of domestic and industrial activities. Contaminants may accumulate during nocturnal inversions, so that levels may reach a maximum during the morning and dissipate rapidly thereafter. Concentrations of pollutants at ground level may show 2 peak periods during a 24-h period, a major maximum in the morning and a minor one in the evening. On the other hand, in areas where photochemical reaction is the predominant factor in the build-up of pollutants under inversion conditions, the peak concentrations are usually attained in the afternoon, after a period of intense solar radiation.

The trend towards the use of automatic instruments in air pollution studies is also the result of the need for information on the following factors:

(*a*) maximum and average levels of contaminants from community and industrial sources, and trends in these levels, for use in judging the effectiveness of control programmes;

(*b*) the level and duration of exposure to contaminants that are responsible for nuisances; irritation of the conjunctiva and other sensitive membranes; offensive odours; reduction in visibility; soiling, corrosion, and other damage to property; and injury to vegetation; and (*c*) the effects of pollutants on health, and methods of protecting the public from exposire to toxic substances at levels or for durations that are harmful.

Such information is also necessary as a basis for the establishment of criteria and standards for control and protective purposes.

Automatic monitoring and sampling instruments [a] use three general methods for sampling gaseous pollutants: (1) continuous sampling, (2) batch-type sampling, and (3) dosimetric sampling by accumulation of the pollutant until a predetermined level is reached. Various principles are employed in monitoring the contaminant gas either directly or by absorption in a suitable reagent solution. Automatic instruments may be based on any of the following analytical techniques: (1) electrolytic conductivity; (2) electrolytic titrimetry; (3) measurement of electrolytic current or potential; (4) colorimetry; (5) turbidimetry; (6) photometry; (7) fluorimetry; and (8) infrared or ultraviolet absorption spectrometry.

Outline descriptions of some of the instruments available for automatic sampling and monitoring of gases are given in Annex 5.

Recent Advances in Instruments for Gas Analysis

In recent years, rapid developments in spectrometry, gas chromatography, and detection methods such as hydrogen flame ionization have made available a variety of new techniques for the continuous analysis of gaseous air pollutants. Such advances have stimulated research in the following areas: the photochemistry of low concentrations of hydrocarbons and nitrogen oxides in air; the identification of the products of such reactions; the effects of such reactions on biological systems; and the monitoring of urban air pollutants. Different workers have greatly improved infrared spectrometric methods by the development of long-path gas cells, double-beam optics, expanded scales, and other modifications.[55,219,240] Both nondispersive and dispersive infrared analysers have been used to measure low concentrations of hydrocarbons and other organic compounds, carbon monoxide, and carbon dioxide. However, the relatively high cost of long-path infrared instruments is a serious disadvantage.

Advances in gas chromatography and the development of hydrogen flame ionization and other ultrasensitive methods of detection, such as β-ionization, activation by clathrate compounds containing tritium, and neutron activation, have provided substitutes for infrared analysers.[158] Hydrocarbons in air may be analysed by means of the flame ionization detector, with or without gas chromatography. Gas chromatographic separation, followed by flame ionization detection, provides an ultrasensitive

[a] Different instruments of this type are described in refs. 33, 40, 50, 115, 126, 152, 176, 190, 195, 254.

means of identifying hydrocarbons and other organic substances and of determining their concentration in the range from parts per million to parts per thousand million. This approach has been utilized by Bellar [31] and by Altshuller [13] in the analysis of hydrocarbons in the Los Angeles atmosphere.[12] A flame ionization hydrocarbon analyser, a gas sampling timer, and a recording potentiometer have been used for the continuous analysis of the concentration of hydrocarbons in the atmosphere and in various parts of a laboratory building in Ottawa, Canada.[75] Hydrocarbons are being measured continuously in a number of cities in the USA as part of the Continuous Air Monitoring Program (CAMP) of the US Public Health Service.

The response to flame ionization is essentially a function of the number of carbon atoms in the molecule. In the combustion of organic compounds, ions of sufficiently long life are generated and collected by a strongly polarized electrode for subsequent identification. This detector has a wide range of response for almost all organic substances. However, its response to organic compounds containing oxygen, sulfur, halogens, and nitrogen is lower than its response to hydrocarbons.

Since inorganic gases and vapours (e.g., hydrogen, nitrogen, oxygen, carbon monoxide, carbon dioxide, and water vapour) have high ionization potentials, they are not ionized at hydrogen flame temperatures. Methods of separating and analysing inorganic gases and hydrocarbons by gas chromatography have been described by Madison [151] and Cvejanovich.[66] Inorganic gases were separated by means of a 5A molecular sieve column, and hydrocarbons by means of gas-liquid partition columns containing squalane on Chromosorb and adiponitrile on Chromosorb, helium being used as the carrier gas in a vapour fractometer.

The degree of refinement that can now be attained in automatic instruments for the continuous analysis of gaseous contaminants in the urban atmosphere is illustrated by the US Continuous Air Monitoring Program referred to above. Since January 1962, this programme has been in operation in Washington, D.C.; Philadelphia, Pa.; New Orleans, La.; Cincinnati, Ohio; Chicago, Ill.; San Francisco, Calif.; St Louis, Mo.; and Denver, Colo. Associated programmes have been in operation in Los Angeles, Calif.; Detroit, Mich.; and New York, N.Y. Concentrations of carbon monoxide, total oxidant, ozone, sulfur dioxide, nitric oxide, nitrogen dioxide, and hydrocarbons are determined at 5-minute intervals. The data for each pollutant are processed on punched tape and analysed by electronic computer to provide short-period averages, peak values, 24-hour means, and other relevant values.

In studies of the response of citrus trees in Southern California to air pollutants (including fluorides, ozone, and peroxyacyl nitrates), an automatic punch-card recording is made of the plant chamber environments and of atmospheric concentrations of nitrogen dioxide, nitric oxide, ozone, total oxidants, fluorides, carbon dioxide, and moisture. The date, time, temperature, and an integrated value for sunlight are also recorded.[112]

Rapid analytical services are necessary to cope with the large number of samples and data that are required for routine measurements and research in air pollution in order to evaluate the effects of pollutants on vegetation and on the health of man and animals. As a result, automatic devices for batch and continuous-sample analysis, based on colorimetric procedures, have been developed. In 1957 Skeggs [229] described a system of this type that incorporated a group of instruments using nonautomatic wet chemical procedures and that automated each step for continuous overall performance. Zaleiko [278] has described Technicon Auto-Analyser instruments that are available for both batch and continuous-sample analysis. These devices contain a dual-beam colorimeter with narrow-band interference filters, a strip chart recorder, and automatic standardization-compensation control. They can be used to analyse air, water, blood, absorbing solutions, body fluids, or prepared tissues. The reagents used in the Technicon Auto-Analyser, and the sensitivities for some typical substances, are listed in Table 4.

TABLE 4. SENSITIVITY OF TECHNICON AUTO-ANALYSER FOR TYPICAL SUBSTANCES, AND REAGENTS USED *

Substance	Range	Reproducibility or deviation	Reagent system
nitrates as NO_3	0.05 mg N/litre	0.003	zinc reduction to nitrite, Griess-Ilosvay method
nitrites as NO_2	0.02-5.0 mg N/litre	± 1.5%	sulfanilic acid, naphthylamine hydrochloride
fluorides	0-2.0 ppm	± 1.0%	alizarin complex, zirconium lake
sulfur dioxide	0-5.0 μg/ml	± 0.5%	bleached rosaniline dye, modified West-Gaeke method
nitrogen dioxide	0.1-5.0 μg/ml	± 0.5%	azo dye
phosphates	0-100 ppm	± 1.0%	molybdovanadic acid
chlorides	0.1-100 μg/ml	1.0%	mercuric thiocyanate
chlorine	0-1.0 ppm	± 1.0%	*o*-tolidine dihydrochloride
cyanides	0-0.1 ppm	± 1.0%	pyridine, benzidine, bromine
silica (water analysis)	0-200 parts per thousand million	± 0.5%	ammonium molybdate, 1-amino-2-naphthol-4-sulfonic acid
tetraethyllead	0-1.0 μg/ml Pb		dithizone reaction

* Data from ref. 278.

CHAPTER 5

Determination of Polycyclic Aromatic Hydrocarbons and Other Carcinogens

Although cancer was recognized as an industrial disease as early as 1775, when chimney sweeps were found to suffer from a high incidence of scrotal cancer, it was not until 1933 that the remarkably potent carcinogen benzo[*a*]pyrene, or 3,4-benzpyrene, was isolated by a research team under Sir Ernest Kennaway from a strongly carcinogenic fraction of coal-tar pitch at the Institute of Cancer Research, the Royal Cancer Hospital, London. This discovery led to a rapid awakening of interest in the isolation and identification of polycyclic hydrocarbons (PAH) and other carcinogenic compounds from many products. It has been found that PAH, including benzo[*a*]pyrene, are widely distributed in the environment. Such compounds have been identified in soot, smoke, fuel combustion products (including the exhaust gases of petrol and diesel engines), tobacco smoke, carbon black, processed rubber, smoked foods, oysters, and many other substances. Within recent years they have also been found, in varying concentrations, in the atmosphere of cities and towns throughout the world.

During the past 20 years, rapid progress has been made in the development of improved analytical procedures for the quantitative separation of PAH and other carcinogens in the atmosphere, tobacco smoke, motor vehicle exhaust gases, and fuel combustion products in general. At the same time a great deal of research has been undertaken, by means of animal experiments and epidemiological studies, to elucidate the relationship between cancer (particularly lung cancer) and PAH in the environment and the smoking habits of different population groups. The rapid increase in lung cancer mortality in many countries within the last 30 years, particularly in the male population, has directed world-wide attention to this problem.

Advances in chromotography and in absorption and fluorescence spectroscopy have led to the development of a number of methods for the separation of at least 25 PAH. Several laboratories in a number of countries now carry out more or less routine estimations of about 8 of these PAH in air samples. Data are available on the monthly and seasonal distribution of these PAH in an increasing number of cities, some of which show relatively high concentrations in the atmosphere. If the usual method of separation is used, more or less quantitative information can be obtained on concentrations of pyrene, fluoranthene, chrysene, benzo[*a*]pyrene, benzo[*e*]pyrene, benzo[*k*]fluoranthene, benz[*a*]anthracene, perylene, benzo[*ghi*]perylene, coronene, and anthanthrene. Sawicki and his co-workers have examined the air of more than 130 urban and nonurban areas in all sections of the USA and found that benzo[*a*]pyrene is universally present, in varying concentration.[215]

Concentrations of PAH in the atmosphere are so small, even in heavily populated cities, that they are reported in micrograms per 1000 m^3 of air sample. In general, the concentration of benzo[*a*]pyrene is relatively low in the summer months, but increases markedly to a maximum during the coldest months, coincident with the increased consumption of fuels for heat and power. Concentrations may vary from less than 1 $\mu g/1000\ m^3$ in relatively clean air to over 100 $\mu g/1000\ m^3$ in the smoky polluted air of a large city. Such trace quantities necessitate the sampling of extremely large quantities of air in order to collect sufficient sample for analysis.

Filtration

The usual procedure for collection of the sample is filtration of air with a high-volume sampler at a rate of 1–1.5 m^3/min, collecting particulate matter on a high-efficiency glass-fibre sheet measuring 20×25 cm. Commins [60] has shown that PAH are not volatilized or otherwise changed significantly when sampled in this manner. However, Commins found that some of the hydrocarbons with lower molecular weights were lost when the sample was kept in the laboratory for 3 weeks after collection. Similar losses occur, leading to unreliable results, if the samples are collected by low-flow-rate pumps over a period of several weeks. For accurate results, the samples should be collected at high flow rates over a short period (not exceeding 24–48 hours) and analysed as soon as possible. In a futher test by Commins, an attempt was made to collect fluorescent PAH in the vapour phase in town air by drawing filtered air through a drechsel bottle containing pure liquid paraffin. (Benzo[*a*]pyrene and many other hydrocarbons fluoresce strongly in this solvent, and its low volatility renders it suitable for use in an experiment of long duration.) No fluorescence could be detected in the liquid paraffin when 81.6 m^3 of air were aspirated over a period of 38 days. Consequently, Commins concluded that no significant amounts of fluorescent hydrocarbons exist in the vapour phase in the atmosphere and that all of the PAH can be collected by filtration of the particulate phase.

Extraction

PAH in filtered particulate matter can be readily extracted with benzene or cyclohexane. Great care must be exercised when these extracts are evaporated to small volume, since some of the lower members of this hydrocarbon series are appreciably volatile. The US National Air Sampling Network uses benzene as the extraction solvent.

The extract is usually reduced to about 50 ml by distillation, and is then evaporated to about 2 ml at room temperature with an electric blower. Extracts of hydrocarbons found in smoke are usually coloured. The hydrocarbons are almost completely stable for several months, if kept in the dark, but when in solution they can be destroyed by ultraviolet light. Consequently, samples that cannot be analysed immediately should be extracted and stored in solvent, protected from light, for later analysis.

Chromatography

The first step in the analysis of a complex mixture of PAH in a solution or extract is adsorption on a chromatographic column of solid adsorbent. Various separation procedures, involving different solid adsorbents and eluting agents, have been used by different investigators. (A review of the literature on the separation and analysis of PAH is given in ref. 213.) Standardized alumina is a convenient adsorbent, and it adsorbs compounds with a greater number of condensed rings (e.g., chrysene) more strongly than those with fewer rings (e.g., phenanthrene). It adsorbs benz[*a*]anthracene more strongly than pyrene, which contains fewer double bonds. These and other differences in adsorption, which are related to structural properties of the molecule, provide a means for selective removal of the compounds from the column by elution with a suitable solvent.

Moore & Katz [157] adjusted the activity of the alumina by drying overnight at 115° C, and then adding 4% (by dry weight) of distilled water. The particulate sample was first extracted with benzene to dissolve the organic fraction and a silica gel column was used to divide the benzene extract into aliphatic, aromatic, and "oxygenated" fractions. The aromatic fraction was then added to the "adjusted" alumina column and the PAH fractions were removed by elution with cyclohexane, mixtures of cyclohexane and benzene, and benzene. The chromatographic technique is essentially that described by Cooper [62] and Commins.[59] Cleary has also described the discrete separation of PAH by the use of long alumina columns of adjusted activity.[53] He employed cyclohexane, followed by cyclohexane with gradually increasing amounts of ether up to a maximum of 30%, as the eluting solvents. In the US Public Health Service method, the adsorbed PAH fraction is eluted with pentane-ether mixtures; it is reported that this method leads to better resolution and chromatographic separation of the constituent hydrocarbons and to sharper ultraviolet absorption peaks with finer structure, and that there is no interference from the solvent.

Absorption and Fluorescence Spectrophotometry

A number of absorption spectrophotometric methods have been used to determine the amounts of hydrocarbons present in chromatographic fractions. The most reliable ultraviolet absorption method is the baseline technique, described in ref. 58. According to Commins,[60] narrow baselines 5–8 mμ in wavelength should be used, to reduce the effect of interference by the spectra of other related hydrocarbons and some unidentified compounds. Selection of the wavelengths on either side of the absorption peak is important.

Absorption spectrophotometry has the advantages that the peak absorption wavelength of a compound is not affected by the presence of other compounds, quenching effects are absent, and the absorption spectra of many pure PAH are available in the literature. Its principal disadvantage is the fact that it is less sensitive than spectral fluorescence methods. Within recent years, as fluorescence spectrophotometers have been perfected and become available, fluorescence spectra have been used by a number of investigators in the analysis of PAH. Such methods are an important step in the identification and analysis of PAH for which excitation and emission spectra are available. The excitation spectrum of a fluorescent hydrocarbon resembles its absorption spectrum. Therefore, if the absorption spectra have been established, the excitation spectrum can serve for the identification of a compound in quantities as small as 10^{-9} g (1 nanogram). For example, benzo[*a*]pyrene has been determined with this degree of sensitivity from fine spectra in octane at –196° C.[a] Furthermore, a mixture of fluorescent compounds can be rapidly detected in a chromatogram spot if excitation at different wavelengths yields different emission spectra. The sensitivity of the excitation and emission spectra of many fluorescent compounds is 10–10 000 times that of their corresponding absorption spectra.[213]

Problems in the Analysis of PAH and Other Carcinogens

The problems involved in the determination of carcinogenic air pollutants were discussed at an international symposium held in Cincinnati, Ohio, in 1961.[214] Moore, Monkman & Katz pointed out a number of such problems, such as the effects of solvents on fluorescence spectra, concentration quench-

[a] Studies of this type are described in refs. 35, 70, 160, 175, 177.

ing, self-absorption, and differences in the intensity of fluorescence of paper-spot chromatograms and that of solutions. Many available polycyclic hydrocarbons have been found to be insufficiently pure for use as reference standards. Since these substances occur in the atmosphere in very low concentrations, extremely sensitive analytical methods are necessary, and compounds of high purity are required for reference standards. Sawicki et al.[217] and other workers have shown that a commercial "pure" hydrocarbon may contain 2–5 other compounds as impurities. For example, so-called "pure" benzo[*a*]pyrene may also contain benzo[*e*]pyrene and perylene. Other impurities present in PAH available as "pure" compounds have been noted by Dubois et al.[72] For this reason, the results of animal experiments on the carcinogenic potency of these substances must be accepted with reserve unless impurities are separated by means of rigorous chromatographic and other analytical techniques.

In past studies, one of the problems involved in the quantitative separation of PAH from air samples has been the difficulty of estimating benzo-[*a*]-pyrene accurately, by either absorption or fluorescence techniques, in the presence of benzo[*k*]fluoranthene. Both compounds occur in the atmosphere and have closely similar absorption and emission spectra. However, Sawicki [217] has shown that the two compounds can be separated by means of thin-layer chromatography.

CHAPTER 6

Rapid Methods for the Identification and Measurement of Air Pollutants

Indicator Tubes and Test Papers

For the preliminary investigation of an air pollution problem, it is desirable to use the simplest and most economical methods available. Sometimes, an inspection of the area involved may suffice to reveal the existence of a problem (e.g., by the condition of exposed surfaces of buildings, the corrosion of wire fences, the discolouration of paint, and the appearance of trees, shrubs, and plants). An initial assessment may also be made by the use of colorimetric indicator tubes or test papers for detecting and identifying common gases and vapours in the atmosphere.

Although direct-reading colorimetric indicators and chemically treated papers have been used for many years to detect and measure toxic substances for industrial hygiene purposes, most of them are not sufficiently sensitive for air pollution studies. The basic principles involved in the use of such methods have been reviewed by Saltzman [204] and by Kitagawa.[136]

Indicator tubes

Indicator-tube methods can detect minute amounts of gas in a short time by means of portable and inexpensive equipment. A pollutant gas causes a change in the colour of a reagent packed in the tube; comparison with colour standards or measurement of the length of discolouration can be used to determine the concentration of the gas. To achieve quantitative results, a known volume of air must be drawn through the tube at a constant rate. Other gases that may be present in the air must be removed, or interference from such gases must be avoided by the use of appropriate reagents in the tube.

It has been widely stated that indicator-tube methods can be used by relatively unskilled personnel. However, although they are simple and rapid, they have many limitations and potential sources of error. Unless the sampling procedure is undertaken or at least supervised by a trained operator, and the results interpreted by an experienced engineer or chemist, the findings may be entirely misleading. Potential sources of error include variations in tube diameters, temperature variations, and inaccurate estimation of stain length or colour intensity. If the colour change in the tube depends upon the rate of reaction, its sensitivity to temperature change is high. For example, with some types of carbon monoxide indicator a correction factor of 2 is required for each 10° C deviation in temperature. The properties of different batches of a given indicating material may vary widely unless close control is exercised during their manufacture (e.g., over the purity and quality of reagents and catalysts, air cleanliness in the factory, and the particle size of the impregnated gel). Each batch of gel must be accurately calibrated. Dynamic calibration methods are to be preferred, particularly with reactive gases such as sulfur dioxide, hydrogen sulfide, oxides of nitrogen, and fluorides, since the use of static methods with such gases might result in large errors owing to losses caused by adsorption. Before use in the field, each batch of indicator tubes should be checked by comparison with a reliable chemical or other method.

The best type of indicator tube contains several layers that are impregnated with different compounds to remove interfering substances. For example, carbon monoxide tubes may contain a layer of gel to remove oxides of nitrogen and hydrocarbons or other organic vapours. In other tubes, the first layer of gel may contain an oxidant that is essential for the indicating material to perform its function properly. For example, tubes for the detection of nitrogen oxides contain a mixture of chromium trioxide and concentrated sulfuric acid to oxidize nitric oxide to nitrogen dioxide, since the indicator responds only to NO_2. The theory of the calibration of indicator tubes is discussed in Annex 6.

Availability

Only a limited number of indicator tubes that are suitable for use in air pollution studies are at present available. Some tubes for detecting carbon monoxide have a sensitivity in the range of about 1.25–12.5 mg/m^3 (1–10 ppm), if they are thermostatically maintained at a temperature of 40° C during a sampling time of 5–10 min. They are available in kits containing a small pump (usually operated by hand) designed to deliver the air sample at the required flow rate. The air volume may be determined by counting the number of strokes of the pump.

Test papers

Test-paper methods may be used for qualitative or semi-quantitative measurement of gaseous pollutants. The paper is exposed to the atmosphere for a known time, and the degree of colour change is measured by comparison with colour standards. Such methods measure the cumulative effect of gaseous pollutants collected on the surface of the paper. If the temperature, humidity, time of exposure, and air velocity are controlled, quantitative results may be obtained. Light reflected by the coloured paper, moving at a constant rate, may be recorded automatically by a photoelectric colorimeter to obtain a permanent record.

Gaseous pollutants such as hydrogen sulfide and hydrogen fluoride may be detected and measured by their extinction effect on the fluorescence of paper impregnated with a fluorescent substance. Cumulative measurement of low concentrations of gases such as sulfur oxides, nitrogen dioxide, and hydrogen chloride can be made by ordinary chemical or instrumental analysis if the paper is appropriately treated after exposure.

Storage

Indicator tubes and test papers have a definite shelf life that is affected by temperature, moisture, impurities, exposure to light, and other factors. Consequently, such products must be carefully designed, produced, and calibrated, and they must be stored in such a way as to minimize the effects of such variables. Storage in opaque moisture-proof containers in a refrigerator is preferable.

Ring Oven Methods

The ring oven technique, introduced by Weisz in 1954 as a microchemical method for separating and identifying inorganic ions in a single drop of solution on filter paper, has been developed rapidly in recent years.[269,271] It appears to be particularly attractive for studies of aerosol pollutants. Samples of airborne particles may be collected by high-volume or paper-tape samplers, electrostatic precipitators, or other appropriate methods. A convenient approach is to collect samples on paper tape and conduct the necessary steps directly on the tape placed on the ring oven.

One drop of sample solution is placed in the centre of a piece of filter paper. Certain substances present in the sample may be precipitated (e.g., sulfides may be precipitated by passing hydrogen sulfide through the paper). The paper is then placed on the ring oven, which consists of a small annular block of aluminium and which is heated to about 110° C. Selected solvents, added by means of a micropipette, are used to wash soluble substances radially outwards from the sample. When the solutions reach the heated oven, the solvents evaporate, leaving a ring on the paper. Soluble substances in this ring, and insoluble substances remaining in the central spot, may then be identified by spot tests.

Ring oven methods for the determination of the following ions have been published:[269,271,274] iron, aluminium, copper, nickel, zinc, lead, vanadium, cobalt, antimony, beryllium, selenium, phosphate, and sulfate. The method will detect these ions in nanogram to microgram quantities. By simple visual matching with a series of standard rings, accuracies of 80–90% can be attained. For greater accuracy, three different concentrations of the unknown are analysed and the rings are compared with three standard rings to obtain a weighted average.[269] In this way, the accuracy can be increased to 92–95%.

Biological Indicators of Air Pollution

It is often possible to assess the nature and magnitude of air pollution by studying its effects on biological indicators such as sensitive species of vegetation. This type of investigation may be undertaken at comparatively low cost and will demonstrate whether more comprehensive methods, involving costly instruments and skilled personnel, are needed.

Biological indicators are particularly useful in detecting low concentrations of toxic gases that are difficult to detect by other simple means. Many gases are much more toxic to plants than to animals, including man, at equivalent concentrations.

Important information on the extent of air pollution may be obtained by comparing leaves collected in the area under study with those collected in adjacent control areas. Sulfur dioxide and fluorides accumulate in the leaves of plants that are exposed to these substances during the growing season. Chemical analysis of leaves collected in the early and late periods of the growing season will provide information on atmospheric concentrations of sulfur and fluorine, and on the distribution of these pollutants throughout the area under study. The rate of accumulation of toxic dusts containing metals and other elements can also be demonstrated in this way.

Effects of gaseous pollutants

The following gases are arranged in order of decreasing toxicity to green plants: hydrogen fluoride, chlorine, sulfur dioxide, ammonia, and hydrogen sulfide.[280] Hydrogen fluoride and sulfur dioxide, which are common air contaminants, produce similar necrotic or bleached areas on the margins and intercostal areas of the leaves of many plants. However, the toxicity of hydrogen fluoride is about two orders of magnitude greater than that of sulfur dioxide. As little as 1 μg of hydrogen fluoride per m^3 of air may be toxic to certain varieties of gladiolus or pine after prolonged exposure. Sulfur dioxide causes adverse effects on the most sensitive plants at the following approximate exposures: 2.9 mg/m^3 for 1 h or 1 mg/m^3 for 8 h; however, 0.29 mg/m^3 may be present throughout the period of growth of a crop of alfalfa, one of the most sensitive species, without causing injury.[122,247]

Thomas [247] has tabulated the relative sensitivity of cultivated and wild plants to injury by sulfur dioxide, hydrogen fluoride, and photochemical "smog" of the Los Angeles type. For example, alfalfa, barley, cotton, and sweet pea are highly sensitive to sulfur dioxide but are quite resistant to hydrogen fluoride. Gladiolus, Chinese apricot, Italian prune, and some varieties of maize are sensitive to hydrogen fluoride but are resistant to sulfur dioxide.

Middleton et al.[154,155] identified air pollutants and studied their distribution in California by examining their effects on plants. They reported characteristic differences in leaf damage caused by ozone and that caused by photochemical oxidants (oxidized hydrocarbons) present in "smog" of the Los Angeles type. Ozone causes the appearance of characteristic dark spots on the upper surfaces of leaves; such damage occurs to pinto beans [155] after exposure to about 0.2 mg/m^3 of ozone for 4 h, and to some varieties of tobacco [57,149] after exposure to about 0.1 mg/m^3 for 4 h. Photochemical "smog" oxidants cause silvering, glazing, or bronzing of the lower surfaces of the leaves of many plants, including chenopodium, meadow grass (Poa), sugar beet, celery, and various kinds of lettuce. Similar silvering or glazing is caused by peroxyacetyl nitrate, one of the important reaction products in photochemical "smog". Sensitive plants can be damaged by exposure to only 0.270 mg/m^3 (0.05 ppm) of peroxyacetyl nitrate for 4–8 hours.[68]

In addition to the silvering effect noted above, oxidants inhibit photosynthesis, suppress growth, and increase the abscission of leaves.

Ethylene is another contaminant that produces characteristic effects on plants—e.g., epinasty of leaves and shoots; the stimulation of abscission cells, resulting in the loss of leaves and flowers; and interference with the normal activity of growth hormones, leading to abnormal flower and plant development. These effects were first discovered by Crocker and his co-workers [64] when marigolds were exposed to 0.06 mg/m^3 and tomato plants to 0.12 mg/m^3 of ethylene. Orchid flowers are highly sensitive to ethylene,[155] their sepals being seriously damaged by as little as 0.006 mg/m^3. Ethylene occurs in the exhaust gas of motor vehicles and in the incomplete combustion products of natural gas and other fuels.

Effects of particulate pollutants

Some particulate pollutants may be highly toxic to plants. Sulfuric acid aerosol, formed by the oxidation of sulfur dioxide, may cause a "pock mark" type of injury when deposited on the upper surface of leaves, particularly when there is fog or when the relative humidity is high. Plants injured in this manner have been found near sources of sulfur dioxide (e.g., smelting plants), near sources of tail gas from sulfuric acid plants, and in urban areas.

Smoke and nonspecific particulate matter are significant pollutants in many urban areas. Such particles are deposited on the leaves of plants, where they clog the stomata and the guard cells, interfering with normal photosynthesis, retarding growth, and eventually leading to the disappearance of susceptible species of vegetation through excessive mortality. Conifers are particularly susceptible to damage by smoke, which causes progressive injury through premature senescence and defoliation.

Darley [67] has described the effects on plants of dust from cement kilns. Such dust, when deposited on leaves in the presence of moisture, kills the palisade and parenchyma cells. Although the dust is deposited mainly on the upper surface of the leaves of deciduous species, it can completely cover the needles of conifers. As a result, the growth of trees is retarded and the yield of agricultural crops is reduced. The affected plants are also rendered more susceptible to infection by fungi. The following types of airborne particulate matter have been reported to be harmful to vegetation: magnesite dust, arsenical compounds, and heavy metals (particularly copper, lead, zinc, iron, nickel, manganese, and mercury) and their compounds.

Lichens

The disappearance of lichens as a result of air pollution, and the use of lichens as indicators of pollution in urban areas of Sweden and other countries, have been described by Skye.[230] For example, it has been found that there is a lichen-free zone in Stockholm that keeps increasing as the city grows in size. The lichens that are least sensitive to the presence of pollutants occur abundantly on various species of tree on the periphery of cities such as Stockholm. Epiphytic lichens are more sensitive to air pollutants, particularly sulfur dioxide, than are other species. Terrestrial lichens appear to be more resistant.

Fenton,[79] in a study of air pollution and its relationship to lichens in Belfast, Northern Ireland, found lichen to be absent from an area extending at least half a mile from the city centre. Measurements of smoke and sulfur dioxide indicate that the air of Belfast is considerably contaminated. Beyond the lichen-free zone, a single species of lichen, *Lecanora conizaeoides*, was found. The extent of growth of this lichen increased regularly and rapidly as the average smoke and SO_2 concentrations decreased with increasing distance from the city centre.

These and earlier studies show that where the air is severely polluted, the lichen flora consists entirely of crustaceous forms and is often limited to only one species beyond the lichen-free zone of highest pollution. With decreasing atmospheric pollution the number of different crustaceous species increases, until a comparatively high percentage of lichen cover is reached. A further decrease in pollution leads to the appearance of foliaceous forms with a more or less corresponding decrease of crustaceous lichens. When air pollution is slight, fruticose forms appear, and eventually—when levels of pollution are extremely low—a normal, competing community of all three forms of lichen is found.

Annex 1

EXPRESSION OF RESULTS

In order to achieve international standardization and comparability, the results of air pollution measurements should be expressed in CGS units. Using this system, the concentration of pollutants is reported in terms of mass per unit volume at a standard or reference temperature and pressure. The standard pressure of 760 mm Hg (1013.25 mb) and the temperature of 0° C should be employed. Normally, the concentration of air pollutants is expressed in terms of milligrams per cubic metre (mg/m^3).

Some substances (e.g., polycyclic aromatic hydrocarbons and analogous compounds that are potentially carcinogenic or highly toxic) occur in the atmosphere in such low concentrations that the recommended mass concentration unit is 1000 or more times the weight of such substances actually existing in a cubic metre of air. In such cases, the concentration may be expressed as micrograms, nanograms, or picograms per cubic metre.

Although correction to standard or reference gas conditions may not be justified for many local purposes, owing to inherent technical errors or the limited accuracy of measurements, it is essential that temperature and pressure conditions be reported in all studies in which comparability and precision are desired.

The use of a great variety of units to express the results of air sampling and analysis has led to much confusion in the literature. Proportional units have been used with no indication of whether the proportion is by weight or by volume, and the temperature and pressure under which the volume was measured have not been reported. Concentrations of gases and vapours of known composition and molecular weight have often been reported in parts per million, per hundred million, or per billion of air by volume. Although a concentration of parts per million by volume can readily be converted into the recommended CGS unit, the term "parts per billion" is confusing, since in some countries (e.g., France, Germany, and the United Kingdom) a billion is regarded as 10^{12} and in others (e.g., the USA) as 10^9.

Computation of the results of atmospheric sampling and analysis is considerably simplified if they are expressed in the units recommended in the accompanying table; the use of these units should also simplify the comparison of different sets of results. The choice of the recommended units or of the alternative or derived units given depends on which is more convenient for reporting data in that it avoids the use of cumbersome decimals.

Where the method of measurement involves an empirical calibration technique, the calibration procedure and derived concentration steps should be clearly indicated. For all practical purposes, the ideal gas law relationship $PV = RT$ is applicable, without appreciable error, for temperature and volume corrections in the range of conditions usually encountered in sampling the atmosphere.

Item	Recommended units	Alternative or derived units
concentration of particulate contaminants (liquid or solid) of known composition	milligrams per cubic metre (mg/m^3)	micrograms per cubic metre ($\mu g/m^3$)
concentration of suspended or airborne particulate matter	milligrams per cubic metre (mg/m^3)	micrograms per cubic metre ($\mu g/m^3$)
concentration of gases or vapours	milligrams per cubic metre (mg/m^3)	micrograms per cubic metre ($\mu g/m^3$) [a, b]
gas volumes	cubic metres (m^3) at standard conditions [b]	
volume emission rates	cubic metres per second (m^3/s)	
mass emission rates	kilograms per second (kg/s)	grams per second (g/s)
velocity	metres per second (m/s)	
air sampling rates	cubic metres per minute (m^3/min) or cubic centimetres per minute (cm^3/min)	litres per minute (litres/min)
temperature	degrees Celsius (°C)	
pressure	millibars (mb) or millimetres of mercury (mm Hg)	
visibility	kilometres (km)	
light transmission	percentage transmittance (%*T*)	
light reflection	percentage reflectance (%*R*)	
particle size	microns (μ) (10^{-6} m)	
wavelength of light	millimicrons ($m\mu$) (10^{-9} m)	ångströms (Å) (10^{-10} m)

NOTE. Time of day should be specified in terms of the 24-hour clock—e.g., 15.00 hours, not 3 p.m.

[a] Parts per million (ppm) may be used as an additional unit, but all results should also be reported in mg/m^3 or $\mu g/m^3$; ppm by volume multiplied by $\frac{\text{molecular weight} \times 10^6}{22\,400}$ yields micrograms of substance per cubic metre of air at standard conditions.

[b] " Standard conditions " means a temperature of 0°C and a pressure of 1013.25 millibars (760 mm Hg).

Annex 2

COLLECTION EFFICIENCY OF THE ELECTROSTATIC PRECIPITATOR

Rose & Wood [197] have shown that the collection efficiency of the electrostatic precipitator (based on turbulent conditions in the gas stream and on the motion of the gas at right angles to the main flow, induced by gas ions streaming from the discharge electrode to the collecting electrode) is given by the expression

$$100(1 - e^{-A_s w/Q})$$

where A_s = total collecting-electrode surface area, m^2;

w = particle drift velocity, m/s; and

Q = volumetric flow rate, m^3/s.

The drift velocity of the particles can be computed from the equation

$$w = \frac{pDE^2}{36 \times 10^7 \pi\mu}\left(1 + \frac{J\lambda}{D}\right)$$

where E = electrical field strength, v/m;

D = particle diameter, m;

μ = absolute viscosity of the gas, poises;

$J = 1.764 + 0.562\, e^{-0.785 D/\lambda}$ = mean free path of gas molecules, m;

and

$p = 3C/(C + 2)$, where C is the dielectric constant in CGS electrostatic units.

The above drift velocity equation is based on the following assumptions: (1) the particle rapidly attains its terminal velocity as it drifts towards the collecting electrode; (2) the fluid friction acting on the particle can be computed from Stokes' law modified by the Cunningham correction for particles having a diameter comparable to or smaller than the mean free path of the gas molecules; and (3) the charge accumulated by a particle rapidly approaches a limiting value.[87]

Annex 3

OPTICAL ABSORBANCE AND REFLECTANCE OF SMOKE STAINS

The absorbance (A) and the reflectance (R) of smoke stains on paper are given by the following expressions:

$$A = \log_{10} (I_0/I) \qquad R = \log_{10} (R_0/R)$$

where I_0 is the intensity of the light transmitted through, and R_0 that of the light reflected by, clean paper; and I is the intensity of the light transmitted through, and R that of the light reflected by, the smoke stain.

In terms of percent transmittance (T) or reflectance, the above expressions reduce to the following forms, assuming the photometer to be adjusted to read 100 $\%T$ for clean paper:

$$A = \log_{10} (100/\%T) = 2-\log_{10} (\%T)$$
$$R = \log_{10} (100/\%R) = 2-\log_{10} (\%R)$$

Various investigators have studied the relationship between the optical density or reflectance of different types of dust collected on filter media and the number, concentration, size, and mass of the particles. Watson & Hounam,[267] in studies of the dust in coal-mines, showed the concentration and absorbance of the particles to be related by the expression $N = aD^b$, where N is the number of coal-dust particles producing a stain on filter-paper, D is the absorbance of the stain, and a and b are constants ($b = 1.5$). Katz et al.,[128] in studies of the atmosphere of Windsor, Ontario, found that a similar curvilinear relation held for the absorbance of a deposit of air-borne matter and the volume of air filtered per unit area of the spot on the filter-paper. The Katz-Sanderson equation may be expressed in the following simplified form:[127]

$$D_L = (L/L_0)^{0.67}$$

where L = linear units of air sampled (volume per unit area of the filter-paper spot);

L_0 = linear units of air sample required to produce unit absorbance (in the same dimensional units as L); and

D_L = absorbance of the spot, derived from measurement of $\%T$.

The relationship between absorbance and the mass concentration or carbon content of spots on paper has also been studied by Katz et al.[128] by means of the equation $M = aD^{1.5}$, where M is the carbon content or mass concentration. The constant a was found to vary with wind and other weather conditions that affect the amount of pollutants swept into the air by atmospheric turbulence. Sullivan,[244] following studies of the dust in the air of Sydney, Australia, proposed the following modification of the Katz-Sanderson equation:

$$M = 3.17\ (100 \times \text{absorbance})^{1.24}/V^{0.67}$$

where V is the volume of the air sample in m³ and M is the mass concentration in μg/m³, for air samples in the range of 0.64–1.27 m³ and for smoke spots 2.54 cm in diameter on filter-paper. Kemény, in measuring air pollution at Pretoria, South Africa, has used the relation $M = aD^{1.67}$ to estimate the mass concentration of suspended matter from the optical absorbance of deposits on filter-paper. He found that values of a ranged from 4.1×10^3 to 48.8×10^3, depending on the wind speed and the season of the year.[134]

Annex 4

NONAUTOMATIC METHODS FOR THE DETERMINATION OF AIR POLLUTANTS

The following sections describe the principles underlying various non-automatic methods for the determination of particulate and gaseous pollutants in air. The sensitivity of each method, as reported by different workers, and substances that have been reported to cause interference, are also listed. It should be noted that the objective of this annex is not to provide a manual of procedures, but merely to give some indication of the complexity of each technique (including the apparatus and reagents that are necessary) and of its suitability for different purposes. For full technical details of the different methods, reference should be made to the publications listed.

ACROLEIN

4-Hexylresorcinol spectrophotometric method [15, 56, 206]

Principle

In the presence of mercuric chloride and trichloroacetic acid, acrolein reacts with 4-hexylresorcinol to produce a blue-purple colour.

Procedure

The air sample is aspirated through a mixture of ethanol, 4-hexylresorcinol, mercuric chloride, and trichloroacetic acid in two midget impingers. When aspiration is complete, the colour is developed by allowing samples of the solution to stand in a 60° water bath. Absorbance is determined spectrophotometrically, and the concentration of acrolein in the sample is read from a calibration curve prepared with a series of dilute acrolein solutions.

Sensitivity

Range: 0.05–25 mg/m^3 (0.02–10 ppm by volume). When two bubblers are used, the absorption efficiency is 95%.

Interference

Most organic pollutants do not cause interference when in ordinary concentrations. Crotonaldehyde and dienes cause minor interference.

ALDEHYDES, ALIPHATIC

See also FORMALDEHYDE, p. 78.

MBTH spectrophotometric method [a]

Principle

In acid solution, aliphatic aldehydes react with 3-methyl-2-benzothiazolone hydrazone (MBTH) to form a blue dye, the intensity of the colour being proportional to the aldehyde concentration. The method can be used for the determination of total water-soluble aliphatic aldehydes.

Procedure

The sample is collected by aspirating air through an aqueous solution of MBTH. The solution is treated with an aqueous solution of sulfamic acid and ferric chloride, and the intensity of the blue colour is measured in a spectrophotometer. The concentration of aliphatic aldehydes, expressed as formaldehyde, is determined from a calibration curve prepared with formaldehyde solutions of known concentration.

---

[a] See refs. 14, 15, 101, 206, 216.

Sensitivity

The lower limit is about 2.68 μg/m³ (2 parts per thousand million), expressed as formaldehyde, using an air sample collected at the rate of 30 litres/h over a 24-h period.

Interference

Aromatic amines, heterocyclic imino compounds, and carbazoles cause interference.

AMMONIA

Colorimetric nesslerization method [115]

Principle

Nessler's reagent reacts with ammonia to produce a yellowish brown colour, whose intensity is proportional to the concentration of ammonia.

Procedure

The sample is collected by aspirating air through dilute sulfuric acid in a standard impinger. The solution is treated with Nessler's reagent, and the intensity of the colour is measured in a universal photometer or in a colorimeter equipped with a blue filter. The concentration of ammonia is determined from a scale prepared with standard ammonium chloride solutions.

ASTM reference distillation method [21]

In the ASTM reference distillation method, the pH of the sample solution is adjusted to 7.4, and the ammonia is distilled into a solution of boric acid. Following nesslerization, the ammonia content is determined colorimetrically as above.

Sensitivity

1 μg per volume of solution tested in the colorimeter.

Interference

Ammonium salts, hydrogen sulfide, and some aliphatic amines cause interference.

ARSENIC

Diethyldithiocarbamate method [a]

Principle

Arsine reacts with a solution of silver diethyldithiocarbamate in pyridine to produce a red colour whose intensity is proportional to the amount of arsenic present.

Procedure

The sample is collected by passing air through an electrostatic precipitator or through sodium hydroxide solution in a standard impinger. It may also be collected on a membrane or Millipore filter. If sodium hydroxide is used, the solution is made neutral to phenolphthalein by the addition of hydrochloric acid, and the arsenic in the sample is reduced to the trivalent form by treatment with stannous chloride and potassium iodide solutions in an all-glass Gutzeit apparatus. Granulated zinc is added to cause the evolution of arsine, which is absorbed in a solution of silver diethyldithiocarbamate in pure pyridine. The absorbance of the solution is measured in a spectrophotometer and the arsenic concentration is determined from a calibration curve.

Sensitivity

(*a*) 0.1 μg of As per sample; (*b*) 0.24 $\mu g/m^3$, using an air sample of 0.85 m^3.

Interference

Hydrogen sulfide interferes, but can be eliminated by the use of a coarse lead acetate plug.

BERYLLIUM

Morin fluorimetric method [b]

Principle

In alkaline solution, Morin reacts with beryllium to yield a complex whose fluorescence is proportional to the beryllium concentration.

Procedure

Air samples are best obtained with a high-volume sampler capable of passing about 0.5 m^3 of air per minute. Whatman No. 42 filter paper, or

[a] See refs. 23, 73, 144, 259. This is the method recommended by the American Conference of Governmental Industrial Hygienists (ref. 16).

[b] See refs. 241, 261, 262.

its equivalent, should be used for filtration; the paper should be substantially free from beryllium or yield negligible blanks. The filtered sample is wet-ashed with nitric and sulfuric acids, and beryllium is extracted with acetylacetone or 8-hydroxyquinoline (oxine). The fluorescence of the Morin complex in alkaline solution is determined in a fluorimeter. The purity of the Morin reagent is a critical factor.

Sensitivity

0.02 μg or less per sample.

Interference

The test is almost specific for beryllium. Copper, iron and zinc interfere, but only if present in large amounts.

Spectrographic method [a]

Procedure

The sample is collected and ashed as for the Morin fluorimetric method. It is then subjected to arc or spark excitation; an internal standard is used and the density of the spectrographic plate is measured with a densitometer.

Sensitivity

0.002–0.05 μg per sample.

Interference

Barium, thallium, alkali metals, and iron interfere.

CADMIUM

Monocolour dithizone method [200, 209]

Principle

Diphenylthiocarbazone (dithizone) reacts with the cadmium ion in alkaline solution to form cadmium dithizonate, which is extracted with chloroform and determined spectrophotometrically. The chloroform extraction, and the use of complexing or chelating agents, remove many metals that might otherwise interfere.

---

[a] See refs. 26, 36, 51, 140, 168, 178, 234.

Procedure

The sample is collected by passing air through an electrostatic precipitator or through dilute nitric acid in either a standard or midget impinger. Dry samples may be collected on filter paper or membrane filters, but should be washed with nitric and sulfuric acids. After treatment with nitric acid, the sample is evaporated to dryness, warmed with nitric acid to dissolve cadmium salts, and diluted with distilled water. The acid in the sample is neutralized with sodium hydroxide in a separating funnel, and the following reagents are added: sodium potassium tartrate, potassium cyanide in strong alkaline solution, hydroxylamine hydrochloride, and dithizone. The cadmium dithizonate is extracted with chloroform [a] and the absorbance of the solution is measured in a spectrophotometer. The concentration of cadmium is read from a standard calibration curve for microgram quantities of cadmium.

Sensitivity

0.1 μg of Cd per sample. The average recovery of Cd is about 95%.

Interference

Thallium causes serious interference, but can be removed by bromine treatment and ether extraction. Other metals do not interfere in ordinary concentrations.

CARBON DISULFIDE

Copper dithiocarbamate method [108]

Principle

In the presence of copper acetate, carbon disulfide reacts with diethylamine or piperidine to form copper dithiocarbamate, the intensity of whose yellowish-brown colour is proportional to the concentration of carbon disulfide.

Procedure

The sample is collected by aspirating air through two absorbers in series. The absorbers contain an alcoholic solution of copper acetate and either diethylamine or piperidine, and are immersed in a bath of ice water. The concentration of carbon disulfide is determined colorimetrically by comparison with standards prepared from solutions of carbon disulfide, or with methyl red solutions.

---

[a] The method recommended by the American Conference of Governmental Industrial Hygienists involves a double extraction procedure (see ref. 200).

Sensitivity

0.5 μg per volume of solution analysed.

Interference

The test is not specific for carbon disulfide in the presence of hydrogen sulfide or thioacetic acid. Hydrogen sulfide can be eliminated by aspirating the air sample through a solution of sodium arsenite and ammonium carbonate.

CARBON MONOXIDE [a]

For the determination of carbon monoxide in air, one of the first three methods described below is usually used.

US National Bureau of Standards indicator-tube method [115, 225]

Principle

Silica gel is impregnated with ammonium molybdate and a solution of palladium in sulfuric acid, forming a palladium silicomolybdate complex. Carbon monoxide converts this complex into molybdenum blue, the intensity of the colour being proportional to the carbon monoxide concentration.

Procedure

A measured volume of air is passed through an indicator tube containing palladium silicomolybdate. The carbon monoxide concentration is determined from the air volume, the time of passage, and the colour change (which varies from faint green to blue).

Sensitivity

(*a*) The lower limit for the development of a green colour is 12.5 mg/m^3 (10 ppm), using an air sample of about 0.25 litre. (*b*) About 1.25 mg/m^3 (1 ppm) may be detected in a 2.5-litre sample at a flow rate of about 100 ml/min.

Interference

Organic substances that can reduce the indicator gel to molybdenum blue interfere.

---

[a] Methods for the determination of CO are reviewed in detail in refs. 113, 115.

Iodine pentoxide method [a]

Principle

At a temperature of 120-150° C, carbon monoxide reacts with iodine pentoxide to form carbon dioxide and iodine. The carbon monoxide present can be calculated either from the amount of iodine or from the amount of carbon dioxide that is formed.

Procedure

A known volume of air is passed over iodine pentoxide, contained in a special tube and heated in an electric furnace. An elaborate purification train must be used to exclude organic vapours, water vapour, carbon dioxide, and particulate matter. The apparatus must be thoroughly purged with pure nitrogen, while the iodine pentoxide tube is heated, for about 5 days before use. The iodine that is liberated is absorbed in potassium iodide solution in a Gomberg trap, and is determined by conventional iodometric titration. If the liberated carbon dioxide is to be determined, the iodine must be removed by absorption in crystalline potassium iodide; the carbon dioxide is absorbed in standard barium hydroxide. The final determination of liberated iodine may be performed colorimetrically or by coulometric measurement of the current generated in a Hersch cell. Above a concentration of 3.75 mg/m^3 (3 ppm) accuracy is high if the liberated iodine is measured coulometrically.

Sensitivity

Lower limit is 1.4–3.5 μg of CO per sample.

Interference

None; the test is highly specific for carbon monoxide.

Infrared spectrometric method

The infrared spectrometric method for determining carbon monoxide, which is relatively accurate, is described on p. 101.

Conversion to methane [74, 182]

Gas chromatography has been used to separate carbon monoxide from the other gaseous constituents of air, followed by reduction of the carbon

[a] See refs. 88, 105, 113, 115, 143.

monoxide to methane by hydrogen and determination of the methane with a flame ionization detector. The method is accurate when the carbon monoxide concentration is greater than 3.75 mg/m³ (3 ppm).

Conversion to carbon dioxide [222]

Procedure

The air is passed through the usual purification and absorption train to remove carbon dioxide and other interfering substances. It is then passed through a tube containing a platinum-wire spiral heated electrically to glowing red heat, converting it into carbon dioxide. The carbon dioxide is absorbed in a known volume of standard barium hydroxide solution, and the excess barium hydroxide is titrated with standard hydrochloric acid.

Sensitivity

(*a*) 0.0007 mg of CO per sample; (*b*) lower limit: 2.5 mg/m³ (about 1 ppm) with an air sample of 0.5 litre.

CHLORIDES

See under HYDROCHLORIC ACID AND CHLORIDES, p. 79.

CHLORINE

o-Tolidine method [a]

Principle

Chlorine reacts with *o*-tolidine to produce a yellow colour, whose intensity is proportional to the chlorine concentration.

Procedure

Air is passed through sodium hydroxide solution in a standard impinger, the temperature being maintained between 20° C and 40° C. The solution is neutralized with sulfuric acid, diluted, and treated with *o*-tolidine in hydrochloric acid. The concentration of chlorine is determined by comparison with colour standards which may be prepared from solutions of copper sulfate and potassium dichromate.

Sensitivity

1 µg in the volume analysed colorimetrically.

---

[a] See refs. 20, 91, 115.

Interference

Oxides of nitrogen, ferric and manganic salts, and other oxidants interfere.

Methyl orange method [138]

Principle

In acid solution, chlorine causes the colour of methyl orange to fade to an extent that depends upon the chlorine concentration.

Procedure

Air is aspirated through a U-shaped absorber containing a porous glass plate and a mixture of methyl orange solution and sulfuric acid. The rate and duration of aspiration depend on the type of test being conducted. The chlorine concentration is determined by comparing the colour of the solution with an artificial colour chart.

Sensitivity

0.2 µg in 2 ml of solution.

Interference

Ozone and oxides of nitrogen interfere if present in concentrations exceeding 0.1 µg/m^3. However, such high concentrations do not normally occur in urban areas.

CHROMIUM TRIOXIDE AND SALTS OF CHROMIC ACID

Diphenylcarbazide colorimetric method [148]

Principle

Chromic acid and its salts react with diphenylcarbazide to produce a red colour whose intensity is a measure of the amount of chromium present.

Procedure

Air is passed through a paper or perchlorvinyl filter, and chromium compounds caught on the filter are dissolved in hot distilled water. A solution of diphenylcarbazide in glacial acetic acid and ethanol is added to a portion of the filtrate. The intensity of the colour is measured photometrically, using a green filter, and the concentration of chromium is determined from a calibration curve prepared from potassium dichromate solutions.

Sensitivity

0.05 μg of CrO_3 per 5 ml of solution.

Interference

High concentrations of iron (>1 mg per 10 ml of solution) and molybdenum (>8 mg per 10 ml of solution) interfere.

FLUORIDES, INORGANIC

Alizarin lake method [a]

Principle

Fluorides bleach an alizarin-thorium or alizarin-complexone-lanthanum lake, the extent of such bleaching being indicative of the fluoride concentration.

Procedure

Air is aspirated through a mixture of sodium hydroxide and hydrogen peroxide solutions in a standard impinger. The sample is partially evaporated to reduce its volume and to destroy organic matter and is then subjected to steam distillation in the presence of perchloric acid in an all-glass (borosilicate glass) apparatus. The bleaching action on one of the lakes noted above is measured spectrophotometrically and the fluoride concentration is determined from a calibration curve prepared with standard sodium fluoride solutions.

Sensitivity

Lower limit: about 0.6 $\mu g/m^3$.

Interference

None; the test is specific for fluorides.

Microdiffusion method [167]

A microdiffusion method has been developed to separate fluoride ions from interfering substances.

Sensitivity

0.2–5.0 μg per ml of sample solution.

---

[a] See refs. 22, 29, 30, 41, 94, 119, 167.

Titanochromotropate method [169]

Principle

The fluoride ion reacts with titanochromotropates to produce a faint rose colour. The intensity of the colour can be enhanced with methylene blue, which acts as an internal colour filter.

Sensitivity

0.3 μg per 1.5 ml of solution.

Interference

The phosphate ion interferes.

FORMALDEHYDE

Chromotropic acid method [15]

Principle

Formaldehyde reacts with chromotropic acid to produce a violet colour whose intensity is proportional to the formaldehyde concentration.

Sensitivity

(*a*) Lower limit: 13 $\mu g/m^3$ (0.01 ppm), using a 60-litre air sample; (*b*) 0.02 μg per 5 ml of solution.

Interference

Phenols, olefins, and aromatic hydrocarbons interfere slightly to yield somewhat low results.

Fuchsin sulfate method [8]

Principle

Aldehydes react with fuchsin sulfate in acid solution to produce a rose-violet colour. The colour produced by aldehydes other than formaldehyde gradually fades out, but that produced by formaldehyde changes to a blue-violet.

Procedure

Air is aspirated through two bubblers in series, each containing distilled water, or through a Rykhter absorber. Fuchsin sulfate reagent and sulfuric acid are added, and the colour is compared with that of a set of standards.

Sensitivity

1 µg per 3 ml of the solution that is tested colorimetrically.

Interference

Other aldehydes do not interfere if their concentration does not exceed 300 µg per 3 ml.

HYDROCHLORIC ACID AND CHLORIDES

Nephelometric method [a]

Principle

Chlorides in the sample are precipitated as silver chloride, which is determined nephelometrically.

Procedure

The sample is collected by passing about 2 m^3 of air through chloride-free distilled water in a midget impinger. Suspended particulate matter is removed from the solution by filtration through Whatman No. 42 filter paper or its equivalent, and the filtrate is treated with a solution of silver nitrate in dilute nitric acid. The silver chloride is determined nephelometrically in a spectrophotometer, and the chloride concentration is read from a calibration curve prepared by treating solutions containing microgram quantities of sodium chloride.

Sensitivity

1 µg of Cl^- per 2 ml of the solution that is tested in the spectrophotometer.

Interference

Other acids, including halogen acids and HCN, cause interference.

Silver nitrate titrimetric method [b]

Principle

The sample solution is titrated with a standard silver nitrate solution, the end-point being determined by the use of an alcoholic solution of dichlorofluorescein as an indicator.

---

[a] See refs. 7, 45, 115, 137.

[b] See refs. 7, 45, 137.

Sensitivity

2 μg of Cl^- in the sample solution.

Interference

Other halogen acids, HCN, and cyanides cause interference.

Alkali or sodium tetraborate titrimetric method

Principle

After precipitation of soluble sulfates and sulfur oxides in the sample, the filtrate is titrated with standard alkali or sodium tetraborate to determine the HCl content.

HYDROGEN CYANIDE [a]

Palladium chelate fluorimetric method [100]

Principle

A chelate compound, potassium bis-(5-sulfoxino)-palladium (II), is formed from palladous chloride, 8-hydroxyquinoline-5-sulfonic acid, and potassium carbonate. Cyanide reacts with this compound, which is not fluorescent, to liberate 8-hydroxyquinoline-5-sulfonic acid, which co-ordinates with the magnesium ion to form a fluorescent chelate. The intensity of fluorescence is a measure of the amount of cyanide present.

Procedure

Air is drawn through potassium hydroxide solution in an impinger. Solutions of potassium bis-(5-sulfoxino)-palladium (II), glycine, and magnesium chloride are added. The intensity of fluorescence is measured, and the concentration of cyanide is read from a calibration curve prepared from standard solutions of potassium cyanide.

Sensitivity

0.02 μg per ml of test solution.

Interference

Fluorides and HF may interfere by quenching the fluorescence.

---

[a] Other methods for the determination of hydrogen cyanide are described in refs. 3, 202.

HYDROGEN SULFIDE

Methylene blue method [a]

Principle

The sulfide ion reacts with *p*-aminodimethylaniline and ferric chloride to yield methylene blue, which is determined spectrophotometrically.

Procedure

Air is passed through an alkaline suspension of cadmium hydroxide (a mixture of cadmium sulfate and sodium hydroxide solutions) in a standard impinger. A solution of *N,N*-dimethyl-*p*-phenylenediamine in sulfuric acid and a solution of ferric chloride are added. The optical density of the methylene blue is determined in a spectrophotometer, and the hydrogen sulfide concentration is read from a calibration curve.

Sensitivity

0.05 μg per ml of test solution.

Interference

Oxidizing agents, including nitrogen oxides and chlorine, interfere. Hydrosulfites may interfere, but are unlikely to be present in air. It has been reported that sulfur dioxide interferes, but it may be removed by passing the air through a prescrubber containing barium acetate and acetic acid.

Silver sulfide method [180]

Principle

Hydrogen sulfide is absorbed in a solution of sodium arsenite and ammonium carbonate. The ammonium sulfate sol that is formed is converted into a silver sulfate sol by the addition of silver nitrate. The silver sulfate is rapidly reduced to silver sulfide, imparting to the solution a yellowish-brown colour whose intensity is a measure of the concentration of hydrogen sulfide.

Procedure

Air is aspirated through two absorbers in series, each containing a solution of sodium arsenite and ammonium carbonate. Silver nitrate is added, and the absorbance of the solution is measured colorimetrically.

---

[a] See refs. 20, 42, 114, 115, 172.

The hydrogen sulfide concentration is read from a calibration curve prepared from standard solutions of sodium thiosulfate treated with silver nitrate.

Sensitivity

1 μg per 5 ml of absorbent solution.

Interference

None; the test is specific for hydrogen sulfide.

LEAD

Mixed-colour dithizone method [a]

Principle

In the pH range of 6–11.5, diphenylthiocarbazone (dithizone) reacts with lead to form a brick-red complex salt, lead dithizonate, which is extracted with chloroform and determined spectrophotometrically.

Procedure

Air is passed through an electrostatic precipitator or through dilute nitric acid in a standard impinger. The samples are ashed in the usual manner with nitric acid, the dry ash is dissolved in dilute nitric acid, and the solution is treated successively with the following reagents: hydroxylamine hydrochloride, an ammoniacal cyanide solution, and a solution of dithizone in chloroform. Treatment with ammoniacal cyanide eliminates interference from all metals except tin (stannous), bismuth, and thallium; however, if the concentration of iron and certain other metals is high in relation to that of lead, a double extraction with the dithizone-chloroform solution is advisable. The absorbance of the solution is measured in the spectrophotometer, and the concentration is read from a calibration curve. All reagents and glassware should be free from lead.

Sensitivity

0.3 μg of lead in the aliquot used for analysis.

---

[a] See refs. 77, 131. This is the method recommended by the American Conference of Governmental Industrial Hygienists and by the US Public Health Service (see ref. 132).

Interference

Heavy metals interfere, but can be removed by the use of ammoniacal cyanide. Tin (stannous), bismuth, and thallium, which interfere and are not removed by ammoniacal cyanide, are unlikely to be present in air in significant concentrations.

Potassium chromate method [181]

Principle

Lead ions react with potassium chromate to form insoluble lead chromate. The degree of turbidity produced is a measure of the concentration of lead.

Procedure

Air is aspirated through ash-free and lead-free filter paper in a metal or plastic holder. The paper is then heated with a mixture of sulfuric and nitric acids until a solid residue is formed. The residue is heated in a muffle furnace at 450–550° C to complete ashing. The ash is dissolved in ammonium acetate solution, and the solution is placed in a colorimeter tube. Solutions of lead nitrate and ammonium acetate are placed in similar tubes. Potassium chromate solution is added to all tubes, and the sample tube is compared with the standard tubes against a dark background. The lead concentration may be determined by means of a calibration scale prepared from standard solutions.[292] The degree of turbidity may be measured more accurately by means of a photometer or a nephelometer.

Sensitivity

(*a*) 0.5 μg of lead in 1 ml of solution; (*b*) 1 μg/m^3 with an air sample of 1 m^3.

Interference

The test is almost specific for lead oxides and salts; however, barium salts cause interference if present.

MANGANESE

Colorimetric method [5]

Principle

Manganese compounds are oxidized to permanganic acid, $HMnO_4$, by ammonium persulfate in the presence of silver nitrate as a catalyst. The intensity of the red-violet colour is a measure of the manganese content.

Procedure

Air is aspirated through a paper or perchlorvinyl filter. The filter is incinerated in a muffle furnace and the ash is treated with a mixture of sulfuric and oxalic acids. The mixture is evaporated to dryness and the residue is dissolved in dilute sulfuric acid. Following centrifugation, silver nitrate and ammonium persulfate solutions are added to an aliquot of the supernatant. The colour is compared with that of standards prepared from solutions of manganous sulfate in dilute sulfuric acid.[297]

Sensitivity

0.8 μg of Mn in 3 ml of solution.

Interference

Chromium interferes. Iron has no effect if its concentration does not exceed that of manganese.

METHANOL

Determination as formaldehyde [15, 115]

Principle

Methanol is oxidized to formaldehyde, which is determined by either the fuchsin sulfite or the chromotropic acid method.

Procedure

Air is aspirated through distilled water in a Rykhter bubbler. A portion of the solution is treated with potassium permanganate solution and sulfuric acid, and the determination is continued as for formaldehyde (see p. 78). If both methanol and formaldehyde are present in air, they can be determined separately by the procedure described in ref. 6.

Sensitivity

0.4 μg/5 ml.

Interference

Interference is caused by other organic compounds that can form formaldehyde under similar conditions. In the absence of such compounds, the test is specific for methanol.

NITRATES

2,4-Xylenol spectrophotometric method [27, 171]

Principle

2,4-xylenol is nitrated, the intensity of the resulting colour being a measure of the nitrate content of the sample.

Procedure

In a high-volume sampler, about 2000 m^3 of air are passed, in a 24-h period, through a glass-fibre or other efficient filter that has been washed free of nitrates and weighed. Only a small aliquot of the sample is needed for nitrate measurement; it is extracted with distilled water under reflux, and the extract is filtered through Whatman No. 1 paper or its equivalent. Sulfuric acid and a solution of 2,4-xylenol in glacial acetic acid are added to the filtrate and the mixture is allowed to stand for 30 min in an oven or water bath at 60° C. The nitrated 2,4-xylenol is removed from other water-soluble coloured substances by extraction with toluene and sodium hydroxide. The absorbance of the extract is measured in a spectrophotometer, and nitrate concentration is determined from a calibration curve prepared from standard potassium nitrate solutions.[289, 291]

Sensitivity

(*a*) 5 μg in 5 ml of water; (*b*) 0.1–10 μg per m^3 of air.

Interference

Chlorides produce low results if present in concentrations greater than 100 μg per ml of prepared sample. Nitrites, if present in the suspended particles in the air, produce high results. Oxidizing agents cause interference.

NITRIC OXIDE

Determination as nitrogen dioxide [a]

Nitric oxide is oxidized to nitrogen dioxide, the amount of which is then determined by means of an acid-permanganate solution. The extent of oxidation in a given apparatus should be checked, since it is sometimes incomplete; the efficiency of conversion may be as low as 60%.

---

[a] See refs. 22, 192, 201, 206, 256.

NITROGEN DIOXIDE

Saltzman modification of Griess-Ilosvay method [a]

Principle

Nitrogen dioxide reacts with *N*-(1-naphthyl)-ethylenediamine dihydrochloride to produce a red-violet colour, whose intensity is measured spectrophotometrically.

Procedure

Air is passed through a solution of *N*-(1-naphthyl)-ethylenediamine dichydrochloride and sulfanilic acid in glacial acetic acid in a fritted glass bubbler. After colour development is complete, the absorbance is measured spectrophotometrically and the nitrogen dioxide concentration is determined from a calibration curve prepared with standard sodium nitrite solutions.

Sensitivity

About 1 μg per ml of absorbing reagent in the concentration range of 0.01–10.0 mg/m^3 (0.005–5 parts per million).

Interference

Ozone and other oxidants cause negligible interference.

OXIDANTS

Neutral buffered KI method [b]

Principle

Oxidants react with potassium iodide to liberate iodine, which is determined spectrophotometrically.

Procedure

Air is passed through a midget impinger or other nozzle-type bubbler containing a solution of potassium iodide in a buffer solution of potassium dihydrogen phosphate and disodium hydrogen phosphate having a pH of 6.8. The absorbance of the solution is measured spectrophotometrically, the iodine concentration being determined from a calibration curve prepared with standard iodine solutions.

---

[a] See refs. 22, 192, 201, 206, 256.

[b] See refs. 43, 46, 206–208.

Sensitivity

0.01 ppm of oxidants (0.021 mg/m^3 expressed as ozone), using a 60-litre air sample. The test will detect ozone, chlorine, hydrogen peroxide, organic peroxides, and other oxidants. However, it will measure only 10% of the true concentration of nitrogen dioxide.

Interference

Sulfur dioxide interferes, but can be eliminated by aspirating the air sample through a glass-fibre filter impregnated with chromium trioxide. Hydrogen sulfide interferes, producing low results. Reducing gases and dusts cause serious interference.

Alkaline KI method [a]

Principle

Ozone and other oxidants react with alkaline potassium iodide to produce a stable product, believed to consist principally of the hypoiodite, from which iodine is liberated by sulfamic and phosphoric acids.

Procedure

Air is passed through a solution of potassium iodide and sodium hydroxide in a midget impinger. After addition of a solution of phosphoric and sulfamic acids, the absorbance of the iodine is measured in the spectrophotometer. The concentration of oxidants, expressed as ozone, is determined with the aid of a calibration curve prepared from standard potassium iodate solutions.

Sensitivity

Lower limit: about 0.042 mg/m^3 (0.02 ppm), expressed as ozone.

Interference

As for neutral buffered KI method.

PHENOLS

p-Aminodimethylaniline sulfate method [b]

Principle

The colour produced when phenol is oxidized in the presence of *p*-aminodimethylaniline sulfate is measured spectrophotometrically.

---

[a] See refs. 22, 43, 206–208, 232.

[b] See refs. 37, 107, 115.

Procedure

Air is aspirated through a solution of sodium carbonate in a fritted-glass bubbler. The solution is transferred to a separating funnel and treated with *p*-aminodimethylaniline sulfate reagent and a solution of calcium hypochlorite. Following extraction with chloroform, absorbance is measured in the spectrophotometer and the phenol concentration is determined from a calibration curve prepared with standard phenol solutions.

Sensitivity

(*a*) 1 μg in the volume of solution analysed; (*b*) about 1.05 μg/m^3 (0.25 parts per thousand million) in an air sample of 1 m^3.

Interference

Interference is caused by other organic compounds that produce the typical colour reaction.

4-Aminoantipyrine method [a]

Principle

In the presence of an oxidant (potassium ferricyanide) and at a high pH, phenols react with 4-aminoantipyrine to produce a red colour; extracted with chloroform, this becomes yellow or orange, and is determined spectrophotometrically.

Procedure

Air is aspirated through sodium hydroxide solution in a standard impinger. Cupric sulfate solution is added and the mixture is transferred to an all-glass still, acidified with phosphoric acid, and distilled. Ammonium chloride solution is added to the distillate and the pH of the solution is adjusted to 10.0 by the addition of ammonium hydroxide solution. Following the addition of 4-aminoantipyrine and potassium ferricyanide solutions, the mixture is extracted with chloroform in a separating funnel. The absorbance of the chloroform extract is measured in a spectrophotometer and the phenol concentration is determined from a calibration curve prepared with phenol.

Sensitivity

0.5 μg in a 5-ml sample that is analysed spectrophotometrically.

---

[a] See refs. 109, 115, 233.

Interference

The test is given by other ring compounds, but under the conditions stated it is relatively specific for phenols.

p-Nitroaniline method [76]

Principle

In acid solution, phenol reacts with diazotized *p*-nitroaniline to produce a complex that is coloured yellow-green in low concentrations and red-brown in high concentrations. The character and intensity of the colour are a measure of the phenol concentration.

Sensitivity

0.2 μg in a 5-ml sample measured colorimetrically.

Interference

Cresols interfere.

SULFATES, PARTICULATE

Barium sulfate turbidimetric method [133, 171]

Principle

Sulfates in the sample are precipitated as barium sulfate, and the resulting turbidity is measured in the spectrophotometer.

Procedure

Air is passed for 24 hours through a sulfate-free glass-fibre or other efficient filter in a high-volume sampler at a rate of about 1.5 m^3/min. Sulfates are extracted from the filter by heating with distilled water under reflux, and the extract is filtered through Whatman No. 1 paper or its equivalent. Hydrochloric acid, a mixture of glycerol and absolute ethanol, and barium chloride crystals are added to the filtrate, and the mixture is allowed to stand at 20–30° C for 40 min. The absorbance is measured in a spectrophotometer, and the sulfate concentration is determined from a calibration curve prepared from sodium sulfate standards.

Sensitivity

(*a*) 2 μg per ml of test solution; (*b*) a range of 1–20 μg/m^3.

SULFUR DIOXIDE

West & Gaeke colorimetric method [a]

Principle

Sulfur dioxide reacts with sodium tetrachloromercurate to form the dichlorosulfitomercurate ion, which reacts with acid-bleached *p*-rosaniline and formaldehyde to produce a red-purple colour.

Procedure

Air is aspirated through a solution of sodium tetrachloromercurate. A solution of *p*-rosaniline hydrochloride in hydrochloric acid and a solution of formaldehyde are added. The absorbance of the solution is measured in a spectrophotometer, and the results are read from a calibration curve prepared with standard sodium metabisulfite solutions.

Sensitivity

0.015–0.6 mg/m^3 (0.005–0.2 ppm), using an air sample of 38.2 litres. The system follows Beer's law up to SO_2 concentrations of about 25 μg per 10 ml of absorbent solution.

Interference

Nitrogen dioxide and ozone interfere if present in concentrations greater than that of sulfur dioxide. If sulfides are present, the precipitate formed in the absorbent must be removed by filtration. Nitrites may cause interference. Nitrogen dioxide interferes, but can be removed by the addition of sulfamic acid to the absorbent. Salts of iron and other heavy metals may interfere, but can be removed by the addition of ethylenediaminetetraacetic acid to the absorbent.

Barium chloranilate colorimetric method [32, 121]

Principle

Solid barium chloranilate reacts with sulfate at a pH of 4 and in alcohol solution to yield the highly coloured (red-purple) acid-chloranilate ion ($HC_6Cl_2O_4^-$).

---

[a] See refs. 106, 115, 164, 170, 179, 245, 272, 273, 281. For modifications necessary for automatic analysis, see refs. 236, 270.

Procedure

Air is aspirated through a solution of hydrogen peroxide. Solid barium chloranilate, 95% ethanol, and a buffer solution of potassium acid phthalate (pH 4) are added. Excess barium chloranilate and precipitated barium sulfate are removed by filtration, and the absorbance of the acid-chloranilate is measured in a spectrophotometer. The sulfur dioxide concentration is determined from a calibration curve prepared with standard potassium sulfate solutions.

Sensitivity

0.15 mg/m^3 (0.05 ppm), using an air sample of 1 m^3. Collection efficiency: about 100%.

Interference

Phosphates, chlorides, and fluorides interfere.

Fuchsin-formaldehyde colorimetric method [a]

Principle

Sulfurous acid is used to develop a red-violet colour in a decolourized fuchsin-formaldehyde solution.

Procedure

Air is aspirated through a solution of sodium hydroxide and glycerol in a standard impinger. A solution of basic fuchsin, sulfuric acid, and formaldehyde is added, and the intensity of the colour is measured in a colorimeter. Standards for comparison are prepared from sodium metabisulfite solutions.

Sensitivity

Lower limit: 0.024 mg/m^3 (0.008 ppm), using 10 ml of solution through which air is aspirated for 30 min at a rate of 5 litres/min.

Interference

Thioacids, thiols (mercaptans), thiosulfates, and inorganic sulfides interfere. However, thiols and thiosulfates can be removed by precipitation with mercuric chloride. Oxides of nitrogen also interfere to some extent.

[a] See refs. 11, 16, 24, 89, 115, 156, 173, 200, 237–239, 258. This method is listed as the recommended procedure in the ACGIH *Manual* (ref. 16), although it has been superseded by the West & Gaeke method (see p. 90).

Stratman colorimetric method [167, 243]

Principle

Sulfur dioxide is reduced to hydrogen sulfide, which reacts with ammonium molybdate to form molybdenum blue. The intensity of the colour is measured photometrically.

Procedure

Sulfur dioxide in the air sample is adsorbed on specially prepared silica gel in a quartz tube. The tube is heated at 500° C to desorb the sulfur dioxide, a stream of hydrogen being passed through the apparatus. The gases are passed through a contact tube containing platinum wire mesh heated to 600° C, and the resulting hydrogen sulfide is passed into a fritted-glass micro-impinger containing a solution of ammonium molybdate, sulfuric acid, and urea. The intensity of the blue colour is measured in a photometer and the sulfur dioxide concentration is read from a calibration curve.

Sensitivity

About 28μg/m^3 when a 30-litre air sample is used. At concentrations below 200 μg/m^3, the standard deviation is $\pm$11.2 μg/m^3.

Interference

Sulfur compounds, including sulfuric acid, interfere. Sulfuric acid mist can be removed by bubbling the air sample through phosphoric acid prior to passing it over the silica gel.

Barium sulfate turbidimetric method [4]

Principle

Sulfur dioxide is oxidized to sulfuric acid, which is determined as barium sulfate.

Procedure

Air is aspirated through a solution of potassium chlorate in sulfate-free distilled water in a U-shaped absorber fitted with a porous glass filter. Barium chloride solution and ethanol are added, and the resultant turbidity is measured by comparison with standards prepared from potassium sulfate and barium chloride solutions.

Sensitivity

2 μg SO_2 per 3 ml of absorbent solution.

Interference

Hydrogen sulfide and sulfuric acid mist interfere.

Acidimetric hydrogen peroxide method [a]

Principle

Sulfur dioxide is trapped in a solution of hydrogen peroxide, which oxidizes it to sulfuric acid, and the acidity is determined by titration with standard alkali.

Procedure

Air is aspirated through hydrogen peroxide solution, adjusted to pH 5, in a bubbler. The acid in the solution is titrated with standard sodium hydroxide, using an indicator composed of bromcresol green and methyl red in methanol.

OECD Working Party modification [167]

The amount of standard sodium carbonate or hydrochloric acid necessary to adjust the volume of hydrogen peroxide used in the test to pH 4.5 is determined. Following aspiration of air through the absorbent, this volume of either sodium carbonate or hydrochloric acid is added. The excess acidity is then titrated with standard sodium carbonate, using a mixed indicator that changes colour from red to grey to blue at pH 4.5.

Sensitivity

The test measures not sulfur dioxide but total gaseous acidity. Lower limit: about 0.03 mg/m^3 (0.01 ppm), expressed as SO_2, using an air sample of about 0.85 m^3. Accuracy: $\pm 10\%$ below a concentration of 0.3 mg/m^3 (expressed as SO_2), increasing with concentration.

Interference

Since the procedure measures total gaseous acidity, acidic gases yield high results. (A downstream filter should be used to remove solid particles and sulfuric acid mist.) Ammonia interferes to yield low results.

[a] See refs. 63, 90, 106, 115, 117, 245, 281.

Barium perchlorate titrimetric hydrogen peroxide method [81, 167]

Principle

Sulfate is directly titrated with barium perchlorate in a solvent other than water, the end-point being indicated by the change of thorin from yellow to pink.

Procedure

Air is aspirated through dilute hydrogen peroxide in a bubbler. Acetone and thorin are added and the pH of the solution is adjusted to about 2.5 by the addition of perchloric acid. The sulfate in the mixture is titrated with standard barium perchlorate solution. The end-point may be determined with greater accuracy by the use of a photometer.

Sensitivity

0.06 mg/m^3 (0.02 ppm), using an air sample of 1 m^3. Accuracy: $\pm 10\%$ if the concentration is greater than 0.12 mg/m^3 and the volume of air sampled is 1 m^3.

Interference

Cations may interfere if not removed by precipitation. Phosphate in particular causes interference, but may be removed by precipitation with magnesium carbonate. Fluorides, in the concentrations normally found in urban air, do not interfere.

Conductimetric hydrogen peroxide method [a]

Procedure

Air is aspirated through hydrogen peroxide solution in a 100-ml bubbler equipped with platinum-plate electrodes (the cell constant should be 0.067 cm). The conductivity is measured with a Wheatstone-bridge conductance meter, the procedure previously having been calibrated against standard sulfuric acid solutions.

Sensitivity

Less than 0.015 mg/m^3 (0.005 ppm) in the range of 0–3 mg/m^3 (0–1 ppm). Reproducibility: within 5%. Accuracy: $\pm 10\%$.

---

[a] See refs. 65, 123, 156, 245, 249, 250, 254, 255.

Interference

Soluble acids, bases, and salts interfere. If aerosols are removed by the use of a dry particle filter, only acidic gases and ammonia interfere.

Nash microconductimetric method [162, 163]

By means of a flexible squeeze-bottle, air is aspirated through dilute hydrogen peroxide in a microconductivity cell. Using one charge of solution, 50 or more readings may be made over 1-min periods.

Sensitivity

Range: 0.3–15 mg/m^3 (0.1–5 ppm).

Determination of SO_2 in the presence of NH_3

If ammonia is present, sulfur dioxide may be determined by either of the following conductimetric methods, as recommended by the OECD Working Party.[167]

Method No. 1

A conductivity cell that has been calibrated with sulfuric acid and ammonium sulfate standards is used to measure the conductivity of two samples of absorbent: a control and the sample through which air has been aspirated. Free sulfuric acid is determined by titration. From calibration curves and from the conductivity data and titration results, the concentration of ammonium sulfate in the exposed sample is determined. The atmospheric concentrations of sulfur dioxide and ammonia can then be calculated.

Method No. 2

The exposed absorbent solution is placed in a conductivity cell that is fitted with platinum electrodes and a magnetic stirrer. The cell is connected to a conductivity bridge, and the sulfate content of the absorbent is determined by titration with standard barium trichloroacetate solution.

Sensitivity

Lower limit: 10 μg sulfate per 10 ml of sample. Accuracy: within 10% when SO_2 concentration is greater than 100 μg/m^3.

Interference

If aerosols are removed by a dry particle filter, acidic gases will be the only source of interference.

Iodine method [a]

Principle

Sulfurous acid is titrated with iodine, using starch indicator.

Procedure

Air is aspirated through sodium hydroxide solution in a standard or midget impinger. The solution is acidified, and the sulfurous acid liberated is titrated with standard iodine solution, using starch indicator.

Sensitivity

0.03 mg/m^3 (0.01 ppm), using an air sample of 0.85 m^3. Reproducibility: 10%.

Interference

Nitrogen dioxide and ozone interfere to yield low results. Hydrogen sulfide and reducing agents interfere to yield high results.

Iodine-thiosulfate method [b]

Principle

Sulfur dioxide is oxidized by a standard solution of iodine, and the excess iodine is titrated with sodium thiosulfate.

Procedure

Air is aspirated through a solution of iodine, potassium iodide, high-grade soluble starch, and sulfuric acid in a standard impinger. The solution is then titrated with standard sodium thiosulfate to a light blue colour.

Sensitivity

0.007 mg/m^3 (0.0024 ppm) in an air sample of 1 m^3.

Interference

Oxidants interfere to yield low results. Hydrogen sulfide and reducing agents interfere to yield high results. Acids, bases, and salts do not interfere.

---

[a] See refs. 106, 115, 231, 245.

[b] See refs. 65, 123, 124, 245.

Polarographic method [a]

Principle

Sulfurous acid is reduced at a dropping mercury electrode, and the sulfur dioxide concentration is determined from a polarogram.

Procedure

Air is aspirated through a solution of sodium hydroxide and glycerol in a standard Greenburg-Smith impinger. Dissolved oxygen is removed from the absorbent solution and from a buffer solution of acetic acid and sodium acetate (pH about 3.9) by purging with nitrogen, and the two solutions are mixed in an electrolysis vessel. Using a polarograph with a dropping mercury electrode, a polarogram is prepared for the range of −0.35 to −1.00 volt. The results are computed from a calibration curve (showing diffusion current plotted against concentration) prepared with standard sodium metabisulfite solutions.

Sensitivity

0.06 mg/m³ (0.02 ppm), using an air sample collected in 75 ml of solution at a rate of 28 litres/min for 30 min. Accuracy: ±10%. Reproducibility: 3%.

Interference

Oxides of nitrogen do not interfere if present in the usual concentrations for urban air. Other sulfur compounds do not interfere.

Lead peroxide cumulative method [b]

Principle

Sulfur dioxide reacts with lead peroxide to produce lead sulfate, which is determined gravimetrically as barium sulfate. The method is designed to measure pollution with sulfur dioxide over a prolonged period.

Procedure

A porcelain cylinder covered with fabric is coated with a paste composed of lead peroxide and gum tragacanth, in such a way that 8 g of lead peroxide cover an area of 100 cm². After the paste is allowed to dry in an atmosphere

[a] See refs. 28, 82, 174, 200. The method is recommended by the American Conference of Governmental Industrial Hygienists for use when oxides of nitrogen are present (see ref. 16).

[b] See refs. 22, 80, 92, 130, 236, 252.

free from sulfur dioxide, the cylinder is exposed to the air in a box fitted with louvres, usually for a period of 1 month. The mixture of peroxide and sulfate is dissolved in a solution of sodium carbonate, and sulfate is precipitated by the addition of barium chloride in dilute hydrochloric acid solution. The barium sulfate is determined by the usual gravimetric procedure, and the concentration of sulfur dioxide is calculated from the result. In British practice, results are usually expressed in terms of milligrams of sulfur trioxide per day per 100 cm^2 of lead peroxide.

Accuracy

Standard deviation from the mean is 7%.

Interference

Sulfur compounds—e.g., hydrogen sulfide—also cause the formation of lead sulfate. Sulfuric acid and particulate sulfates also interfere in this way.

SULFURIC ACID MIST

Sodium hydroxide titrimetric method [a]

Procedure

Suspended matter, including sulfuric acid mist, is collected on filter paper or a Millipore filter in the same way as in the filtration of smoke. The filter is macerated in distilled water and the resulting solution of acid is titrated with standard sodium hydroxide solution, the end-point being determined by means of a pH meter.

Sensitivity

The sensitivity of the method depends upon the sampling rate and the volume of the air sample. With a sample of 3 m^3, the lower sensitivity limit is about 16 $\mu g/m^3$.

Interference

Aerosols of other soluble electrolytes may cause interference. Ammonia also interferes, producing low results.

---

[a] See refs. 61, 97, 150, 281. Samples may also be collected by means of a cascade impactor, by impaction at high velocity (refs. 248, 263), or by electrostatic precipitation (ref. 221).

Sodium tetraborate titrimetric method (Commins) [61]

Procedure

The sample is collected on Whatman No. 1 filter paper or its equivalent. The acid and soluble salts are removed by washing with distilled or deionized water and the free acid is titrated with standard sodium tetraborate solution to pH 7.0. Alternatively, a known small excess of sodium tetraborate may be added and the excess back-titrated with standard sulfuric acid.[191]

Sensitivity

Less than 1μg/m^3 with an air sample of about 10 m^3 sampled at a rate of 20 litres/min.

Interference

Soluble bases (e.g., metallic oxides and hydroxides) interfere. Ammonia also interferes, but can be determined separately by nesslerization.

Other methods

Sulfuric acid and sulfates can be determined nephelometrically, the sensitivity of the method being about 1 μg/m^3. Other methods are discussed in ref. 167.

SULFURIC ACID MIST AND SULFUR DIOXIDE (COMBINED DETERMINATION)[9]

Procedure

Air is aspirated through a U-shaped absorber containing a fine porous glass plate and a solution of sodium hydroxide and glycerol. The solution is then divided into two parts. In one part, sulfur dioxide is determined by the colorimetric fuchsin-formaldehyde method (see p. 000). Sulfur dioxide is driven out of the other part by aspiration with air; barium chloride is added, and the resultant turbidity is measured nephelometrically. The concentration of sulfuric acid is determined from a calibration scale.

Sensitivity

0.5 μg of H_2SO_4 in 2 ml of absorbent solution.

Interference

Particulate sulfates, if present in the air sample, will cause high results.

Annex 5

AUTOMATIC SAMPLING AND MONITORING INSTRUMENTS

Some automatic instruments for the sampling of air and the monitoring of gaseous contaminants are briefly described in the following sections. For further details of the procedures and instruments, reference should be made to the publications listed.

GASES THAT CAN BE ABSORBED IN, OR THAT REACT WITH, REAGENT SOLUTIONS

General methods [a]

Sequential or multiple samplers, such as automatic impingers or sequence bubblers, are used to collect 1-h or 2-h samples. The sampling cycle is actuated by a clock mechanism and air is drawn through the device by a motor-driven pump, equipped with a filter to remove particulate matter; an air flow meter is also included. Following collection, samples may be sent to a laboratory for analysis. However, some devices may treat the samples and measure and record the results. Such devices automatically dispense and transfer solutions and carry out titrimetric, colorimetric, spectrophotometric, or conductimetric procedures, the different steps being co-ordinated by servo-mechanisms. The sampling rates of such devices may be 0–10 or 0–60 litres/min.

[a] See refs. 52, 118, 156, 275.

Many of the methods used are nonspecific, and the sensitivity and precision of the devices must be calibrated for a given sampling site.

Colorimetric methods [1]

In dosimeter-type instruments, an air sample of known volume is passed through a reagent solution in a colorimeter cell. When a preselected photocell output (corresponding to a given degree of colour development and thus a given concentration of absorbed pollutant) is reached, the time is stamped on a paper strip chart and the colorimeter cell is emptied and refilled with reagent.

CARBON MONOXIDE

Infrared spectrometric methods [116]

Principle

Since it depends on a fundamental physical property of the molecule, nondispersive infrared spectrometry is an excellent technique for measuring the concentration of trace gases in automatic instruments. However, it is relatively costly. In this type of analysis, the infrared absorption bands having wavelengths of 1–15 microns are usually used. A pressurized infrared spectrometer has been devised to monitor carbon monoxide in the range found in the atmosphere. The instrument contains sample and comparison cells, a radiation source (consisting of two heated Nichrome wires), an infrared optical system, a detector, an electronic amplification system, and a recorder. One infrared beam passes through the sample cell and the other passes through the comparison cell; both emergent beams converge on the detector. When the radiation in the two beams is equal in intensity, no output signal is set up in the detector. However, when an air sample containing carbon monoxide is introduced into the sample cell, the intensity of radiation in the beam from this cell is reduced; the resultant imbalance in the radiation of the two beams sets up a signal that is amplified and recorded.

Errors caused by water vapour in the air sample can be almost eliminated by maintaining the relative humidity of the sample and of the comparison cell at about 100%, and by calibrating the instrument accordingly.

Sensitivity

0–25 mg/m^3 (0–20 ppm) or greater.

Interference

None; the method is highly specific.

FLUORIDES

See HYDROGEN FLUORIDE, p. 103.

HYDROCARBONS

Infrared spectrometric methods [116, 146]

Principle

Nondispersive infrared spectrometers similar to those used for the determination of carbon monoxide may also be used to determine hydrocarbons in terms of hexane.

Sensitivity

0–15.3 mg/m³ (0–4 ppm), expressed as hexane.

HYDROGEN CYANIDE AND HYDROGEN SULFIDE

Potentiometric method [190, 242]

Principle

Hydrogen sulfide and hydrogen cyanide in the air sample are absorbed in sodium carbonate solution, which then passes into a cell containing a silver electrode and a saturated calomel reference electrode. The half-cell potential of the silver electrode depends on the concentration of silver ions, which in turn is influenced by the presence of cyanide and sulfide ions. The potential of the silver electrode, measured against that of the calomel electrode, is proportional to the concentration of cyanide and sulfide ions.

Sensitivity

1.3 $\mu g/m^3$ (1 part per thousand million) for HCN, with a reproducibility of 10%. The method is not sufficiently sensitive for the measurement of low concentrations of H_2S.

Interference

Ionizable and acidic gases that are soluble in the absorbent solution interfere.

HYDROGEN FLUORIDE

Electrolytic galvanimetric method [190, 268]

Principle

Hydrogen fluoride in the air sample is absorbed in a solution of acetic acid, which then passes into a galvanic cell. The increase in current flow that occurs between aluminium and platinum electrodes is proportional to the fluoride concentration.

Sensitivity

Range: 0.1–10 mg/m^3.

Interference

Ionizable and acidic gases that are soluble in the absorbent solution interfere.

Fluorimetric method [48, 253]

Principle

The air sample is divided into two measured streams, from one of which fluorides are removed. The two streams are passed through adjacent areas of filter-paper tape that is impregnated with a solution of magnesium oxinate in methanol and illuminated by ultraviolet light having a wavelength of about 3650 ångströms. Fluorescent light passes through a 4500-ångström filter and is measured by two photomultiplier tubes whose output currents are continuously recorded on a strip chart. Gaseous fluorides present in the stream of unaltered air cause marked quenching of fluorescence.

Sensitivity

Performance is reproducible within 10% in the range of 0.2–10 μg/m^3.

Interference

None; the test is highly specific for gaseous fluorides.

HYDROGEN SULFIDE

See also HYDROGEN CYANIDE AND HYDROGEN SULFIDE, p. 102; *and* SULFUR DIOXIDE, electrolytic titrimetric method, p. 112.

Thomas Autometer conductrimetric method [246, 254]

Principle

Oxidizable sulfur compounds are converted into sulfur dioxide by passage over a platinum wire heated electrically to 550° C. The sulfur dioxide is absorbed in dilute, acidulated hydrogen peroxide solution, and the resultant increase in conductivity is measured.

Sensitivity

About 15 $\mu g/m^3$ (1 part per hundred million) of H_2S.

Interference

The method measures total oxidizable sulfur compounds and therefore is not specific for H_2S.

NITRIC OXIDE

Griess-Ilosvay colorimetric method [34]

Principle

Nitric oxide in the air sample is oxidized photochemically by light from a high-pressure mercury vapour lamp in the presence of 0.05% of butadiene. The nitrogen dioxide is absorbed in Griess-Ilosvay reagent and the intensity of the coloration is measured.

Sensitivity

About 0.014 mg/m^3 (0.01 ppm).

NITRIC OXIDE AND NITROGEN DIOXIDE (COMBINED DETERMINATION)

Griess-Ilosvay colorimetric method [201, 256]

Principle

The air sample is divided into two measured streams. Nitric oxide in one stream is converted into nitrogen dioxide by passing through an acidified solution of potassium permanganate. Both streams are passed into counter-current absorbers, where they are absorbed in solutions of sulfanilic acid, *N*-(1-naphthyl)-ethylenediamine dihydrochloride, and acetic acid. Both solutions pass into colorimeter cells for the measurement of absorbance. Readings obtained from the solution of unaltered sample give the nitrogen dioxide concentration. The difference between this value and that for the permanganate-treated sample gives the nitric oxide concentration. The efficiency of conversion from NO to NO_2 varies from 70% to more than 95%, depending upon the design of the bubbler.

Interference

Ozone causes slight interference in concentrations up to five times that of NO_2. Sulfur dioxide has no effect in concentrations up to ten times that of NO_2.

NITRIC OXIDE, NITROGEN DIOXIDE, OZONE, AND OTHER OXIDANTS (COMBINED DETERMINATION) [a]

Principle

An automatic instrument containing three recording colorimeters with associated air-sampling and liquid-absorption systems is used. Nitric oxide and nitrogen dioxide are determined as described above, and oxidants are measured by the liberation of iodine from a buffered, neutral solution of potassium iodide. The light transmission of the yellow solution that is formed is measured in a double-cell colorimeter and recorded continuously on a strip chart. The absorbing solution is re-used after being regenerated by passage through a bed of activated carbon.

Sensitivity

Oxidants: 0–2 mg/m^3 (0–1 ppm), in terms of ozone. Absolute accuracy may be established by calibration against known concentrations of ozone.[22]

Interference

Reducing substances, such as SO_2 and H_2S, cause serious interference. However, SO_2 may be removed by passing the sample through glass-fibre paper impregnated with chromium trioxide.[208]

---

[a] See refs. 43, 145, 208.

NITROGEN DIOXIDE

For the combined determination of nitric oxide and nitrogen dioxide, see p. 104; *for the combined determination of sulfur dioxide and nitrogen dioxide, see* p. 113.

Sulfanilamide diazotization method [118]

Principle

The sample is absorbed in an alkaline solution (sodium hydroxide and butanol), and hydrogen peroxide is added to eliminate interference of sulfur dioxide by converting it into sulfate. Nitrogen dioxide is determined by the diazotization of sulfanilamide in the presence of phosphoric acid, at a pH of less than 2, followed by coupling with *N*-(1-naphthyl)-ethylenediamine dihydrochloride. The absorbance of the solution is determined photometrically.

Sensitivity

4 μg per 50 ml of test solution; 0.08 mg/m³ using a 50-litre air sample.

Interference

Sulfur dioxide interferes, but is eliminated as noted above.

OXIDANTS

See also OZONE, p. 107. *For combined determination of nitric oxide, nitrogen dioxide, ozone, and other oxidants, see* p. 105.

Neutral buffered KI method [a]

Principle

The sample is passed through a neutral buffered solution of potassium iodide, the colour of which is then measured in a recording colorimeter at a wavelength of 3600 ångströms, or a wavelength range of 3400–3800 ångströms, the light being obtained from a mercury-vapour lamp equipped with two ultraviolet filters. The instrument may be dynamically calibrated with an ozone-air mixture. One portion of the mixture is analysed as above; the other is passed through a faintly blue solution of iodine, potassium iodide, and starch, which is then titrated with standard sodium thiosulfate solution.

---

[a] See refs. 22, 52, 191.

Sensitivity

0.02 mg/m³ (expressed as ozone) with a 30–min sample.

Interference

Ozone, nitrogen dioxide, chlorine, and some organic peroxides liberate iodine and give high results.

Phenolphthalein method [98, 195]

Principle

The method is based on the oxidation of phenolphthalin to phenolphthalein and is used in automatic instruments for batch-type sampling and analysis. The sample is aspirated through a solution of phenolphthalin and cupric sulfate, which in the presence of oxidants turns red-purple. A reagent containing ferrous thiocyanate may also be used. Absorbance is measured before and after sampling, and results are recorded on a strip chart. Instruments calibrated with hydrogen peroxide yield results twice as high as those obtained from instruments calibrated with ozone.

Sensitivity

0–2 mg/m³ (0–1 ppm), expressed as ozone.

Interference

Sulfur dioxide, nitrogen dioxide, peroxides, and other substances able to reduce phenolphthalein interfere.

OZONE

For the combined determination of nitric oxide, nitrogen dioxide, ozone, and other oxidants, see p. 105.

Electrolytic titrimetric method [186, 190]

Principle

The apparatus consists of a reaction cell equipped with a pair of " production " electrodes and a pair of " sensor " electrodes. An electric current passing through the production electrodes liberates iodine from a potassium iodide solution that flows continuously into the cell. The iodine is maintained at a constant, very low concentration (the " zero " level of the instrument) and is prevented from volatilizing by the controlled addition of sodium thiosulfate.

The air sample is passed at a measured rate through the potassium iodide solution, from which ozone, if present, quantitatively liberates iodine. The resultant increase in iodine concentration causes a corresponding increase in the current flowing through the sensor electrodes. This current, in turn, operates a servo system that reduces the current flowing through the production electrodes, thus causing the iodine concentration to return to its low " zero " level. The ozone concentration is determined from the extent to which the current in the production electrodes must be reduced in order to re-establish the " zero " level.

In practice, two separate reaction cells are used, and the air sample is divided into two equal streams. The inlet of each cell is equipped with a device that can heat the air to a temperature of 300° C. Both heaters are turned on, thus dissociating any ozone in the air sample, and the currents flowing through the production electrodes in the two cells are balanced. One of the heaters is then turned off, and the amount of ozone in the unheated air sample is determined from the difference between the currents in the production electrodes of the two cells. The results are automatically recorded on a strip chart.

Sensitivity

Full-scale range: 0–2 mg/m³ (0–1 ppm); the minimum detectable level is 0.02 mg/m³.

Interference

The only substances that interfere are sulfur compounds that do not oxidize completely at 300° C.

Coulometric titrimetric method [a]

Principle

The method is based on continuous titration of the iodine liberated from a neutral, buffered solution of potassium iodide. The iodine that is liberated causes a current to flow in a cell equipped with platinum electrodes; this current operates a relay that controls the addition of standard sodium thiosulfate from an automatic burette.

Sensitivity

One commercial instrument based on this principle records a sensor current of 1 μA with an ozone concentration of 0.1 mg/m³ (0.05 ppm).

[a] See refs. 38, 83, 260, 264.

Reaction with nitric oxide [207]

Principle

The air sample is mixed with nitric oxide in pure nitrogen at low pressure in a stainless steel cylinder. Ozone in the sample rapidly oxidizes the nitric oxide to nitrogen dioxide. The nitrogen dioxide is absorbed in *N*-(1-naphthyl)-ethylenediaminetetraacetic acid, and the resultant violet colour is estimated in a colorimeter. Nitrogen dioxide in a separate air sample is determined at the same time, and the difference between the two values gives the ozone concentration. The efficiency of conversion of ozone is 95% or higher.

Sensitivity

0.02 mg/m^3 (expressed as ozone) with an air sample of about 30 litres.

Interference

Organic oxidants do not interfere.

Ultraviolet spectrometric method [189]

Principle

A high-pressure mercury vapour lamp, placed at the focus of a spherical mirror, transmits a beam of ultraviolet radiation over a path about 100 m in length. The beam is received by a prism spectrometer, and the ozone concentration in the atmosphere is determined from the ratio of the transmitted to the incident radiation. (The lines at 265, 280 and 313 mμ are examined.) This method has the advantage of avoiding any handling or disturbance of the air.

Sensitivity

Less than 0.04 mg/m^3 (0.02 ppm) to more than 1.0 mg/m^3 (0.5 ppm).

Interference

None; the method is highly specific.

Ultraviolet photometric method

Principle

The ultraviolet photometer contains two absorption cells and a source of 2537-ångström radiation. The air sample containing ozone is passed through one cell and a similar sample from which ozone has been removed by catalytic decomposition is passed through the other cell. The ozone concentration is calculated from the difference in photometer readings for the two cells. The device is operated in a range in which optical density varies inversely with transmittance, and a reduction of 0.4% in transmittance causes a full-scale deflection (2 mg/m^3, or 100 parts per hundred million, of ozone).

Sensitivity

0–2 mg/m^3 (0–100 parts per hundred million).

Interference

None; the test is highly specific.

Fluorimetric method

A fluorimetric method for the determination of ozone has been developed by Regener.[187]

SULFUR DIOXIDE

p-Rosaniline method [a]

Principle

The air sample is passed through a counter-current absorber, where sulfur dioxide is absorbed in a solution of *p*-rosaniline hydrochloride, hydrochloric acid, and formaldehyde adjusted to pH 1.5. The absorbance of the solution is measured and continuously recorded by a photoelectric colorimeter that is calibrated against standard sodium metabisulfite solutions. The absorption efficiency is 96%.

Sensitivity

The instrument responds to variations of 0.002 μg of SO_2 per ml of absorbent solution. Over a 4-min period, the lower limit is 26.4 μg/m^3. Over a 30-min period, 3.5 μg/m^3 may be detected.

Interference

Nitrogen dioxide does not interfere in concentrations up to 2 mg/m^3 (1 ppm); higher concentrations reduce the optical density of the reagent. Ozone in the gas phase does not interfere in concentrations up to about 6 mg/m^3 (3 ppm).

[a] See refs. 102, 110, 156, 170.

Electrolytic conductimetric method [a]

Principle

Sulfur dioxide is absorbed in high-grade distilled water containing small amounts of sulfuric acid and hydrogen peroxide. The gas is oxidized to sulfuric acid and the conductivity of the solution is measured, using platinum electrodes, by a Wheatstone bridge device with a range of 125–50 000 ohms. Both "cumulative" and "instantaneous" types of analyser are available. The absorption efficiency is 98–100%.

Sensitivity

Lower limit: about 28 $\mu g/m^3$. Upper limit: 9–15 mg/m^3 or more, depending upon the volumes of absorbent solution and gas sample.

Interference

The method is not specific for sulfur dioxide, since conductivity is a property of all ionic solutions. Carbon dioxide in the atmosphere does not interfere.

Davis electrolytic conductimetric method [b]

Principle

Air is passed through water that has been purified by passage through an ion-exchange resin, and the acidity of the water is measured conductimetrically. (In many communities, sulfur dioxide is the major ion-forming contaminant. The results may be expressed in terms of the total acidity of the air.) The ion-exchange resin contains a substance that indicates when it is exhausted. In addition to the ion-exchange unit, the device contains a pump, a flow-control system, conductivity cells, and a recorder.

Sensitivity

Full-scale range: 0–3 mg/m^3 (0–0.95 ppm). The range may be adjusted electrically or by control of the flow rates of the air and water. For a full-scale range of 1.5 mg/m^3 (0.5 ppm), the lower limit may be about 50 $\mu g/m^3$ (0.02 ppm).

[a] See refs. 65, 95, 156, 255. The method has been adopted as standard by the American Society for Testing and Materials (see ref. 22).

[b] The method is used in the Davis Electroconductivity Analyzer, manufactured by Davis Instruments, Newark, N.J., USA. (See ref. 115, p. 394.)

Interference

Aerosols and nitrogen dioxide cause little interference in the " instantaneous " type of absorber. Sulfuric acid aerosol usually passes through the absorbers readily and causes only minor interference, unless the relative humidity exceeds 85% and the aerosol droplets become relatively large. Variations in the atmospheric concentration of carbon dioxide may cause interference, since the pH of the water is not adjusted by the addition of acid.

Iodometric method [a]

Principle

The air sample is aspirated through a bubbler containing a solution of iodine, potassium iodide, and starch. The change in the light transmittance of the solution resulting from the absorption of sulfur dioxide is measured photoelectrically. The device aspirates air through two bubblers alternately, every 2 min. Light transmittance is measured in one tube while air is aspirated through the other. The bubblers are automatically drained and replenished with solution after eight periods of aspiration.

Sensitivity

About 7 $\mu g/m^3$ in a sample of 1 m^3. Upper limit: 2860 $\mu g/m^3$.

Interference

Interference is caused by hydrogen sulfide, nitrogen dioxide, ozone, and other oxidants, and by other substances that either reduce iodine (giving high results) or oxidize iodide (giving low results). The effect of nitrogen dioxide is slight, owing to the fact that it is not well absorbed in the type of bubbler that is used. Sulfuric acid, salts, and bases (e.g., ammonia and calcium hydroxide) do not interfere.

Electrolytic titrimetric method [b]

Principle

The apparatus consists of a two-chambered cell containing a solution of potassium bromide. Bromine is liberated by a current that flows through two generating electrodes. Sulfur dioxide reacts with the bromine, momentarily reducing its concentration and causing a change in the current

---

[a] See refs. 123, 249–251.

[b] See refs. 22, 223, 265.

that flows through a pair of sensor electrodes. The latter current controls (through an amplifier) the current in the generating electrodes. Thus, when the bromine concentration drops as a result of reaction with sulfur dioxide, the current in the generating electrodes increases and more bromine is produced, so that the bromine concentration is maintained constant. Consequently, the amplifier output current is proportional to the sulfur dioxide concentration.

Sensitivity

0.15–0.3 mg/m^3 (0.05–0.1 ppm), depending upon whether the full-scale range is 15 or 30 mg/m^3 (5 or 10 ppm). The accuracy of the method is 5% over short periods; when calibration is carried out monthly, the accuracy is 10%.

Interference

Hydrogen sulfide, thiols (mercaptans), and organic disulfides yield high results. Ozone, nitrogen dioxide, and halogens yield low results.

SULFUR DIOXIDE AND NITROGEN DIOXIDE (COMBINED DETERMINATION)

Principle

Air from the same source is passed through three automatic impingers. One impinger contains hydrogen peroxide solution (for conductimetric determination of sulfur dioxide); one contains sodium hydroxide and glycerol (for colorimetric determination of sulfite by the fuchsin-formaldehyde method)[156, 173, 258] and one contains sulfanilic acid, acetic acid, and *N*-(1-naphthyl)-ethylenediamine dihydrochloride (for determination of nitrogen dioxide).[201]

Sensitivity

Lower limit for sulfur dioxide: 0.1 μg per ml of fuchsin reagent, or 11 μg per m^3 of air sample. Range for nitrogen dioxide: about 1.0–2.0 mg/m^3 (0.5–1.0 ppm). About 80% of the nitrogen dioxide in an air sample is absorbed in one fritted-glass absorber.

Interference

Nitrogen dioxide interferes with the fuchsin-formaldehyde method so that results for sulfur dioxide obtained by this method are consistently lower than those obtained by the conductimetric method.

Annex 6

CALIBRATION OF INDICATOR TUBES

Saltzman [204] has discussed the theoretical basis of the calibration of indicator tubes and has analysed the following variables that can affect the length of the stain: the concentration of gas, the volume of the air sample, the sampling flow rate, the temperature, the pressure, and tube construction features. In the usual type of detector tube, equilibrium is not attained between the sorbed gas and the indicating gel since the flow rate is relatively high and the sampling period is short. Theoretical analysis shows that the stain length is proportional to the logarithm of the product of gas concentration and sample volume, as follows:

$$L/H = \ln(CV) + \ln(k/H) \tag{1}$$

where L = length of stain, cm;
C = gas concentration, ppm;
V = air sample volume, cm^3;
k = a constant for a given type of indicator tube and test gas; and
H = height of a mass transfer unit, cm.

The factor H varies as F^x, where F is the sampling flow rate and x has a value between 0.5 and 1.0, depending on the nature of the process that limits the kinetic rate of sorption. (The limiting process may be diffusion of the test gas through a stagnant gas film surrounding the gel granules, the rate of chemical reaction at the surface, or diffusion into the pores of the gel particles.) If the indicator tube follows equation (1) above, a curve of stain length plotted against log CV (for a constant, fixed flow rate) will be a straight line of slope H.

In some types of tube, the reaction between the gas and the indicating gel is sufficiently rapid to attain equilibrium. Consequently, a stoichiometric relation exists between the volume of gel and that of absorbed gas. For the simplest case, the stain length is proportional to CV, or

$$L = kCV \tag{2}$$

The concentration exponent may differ from unity if adsorption is the controlling factor, according to the relation

$$L = k_2 C^{(1-n)} V \tag{3}$$

The factor n in equation (3) has the same value as in the Freundlich isotherm equation for equilibrium adsorption. If n has the value of unity, the stain length is independent of the concentration and is proportional only to the sample volume. Obviously, such an indicator tube would have no practical value. Equilibrium conditions may be assumed for a given type of indicator tube if stain lengths are directly proportional to air sample volume (at constant concentration) and are not affected by the sampling flow rate.

References

1. Adams, D. F., Dana, H. J. & Koppe, R. K. (1959) *J. Air Pollut. Control Ass.*, **9,** 160
2. Air Pollution Control Association (1955) *J. Air Pollut. Control Ass.*, **5,** 176
3. Aldridge, W. J. (1944) *Analyst*, **69,** 262
4. Alekseeva, M. V. (1934) *Z. prikl. Him.*, No. 4, p. 616
5. Alekseeva, M. V. (1963) [*The determination of manganese aerosol.*] In: Alekseeva, M. V., ed. [*The determination of atmospheric pollutants*], Moscow, Medgiz, p. 105
6. Alekseeva, M. V. (1963) [*The separate determination of formaldehyde and methyl alcohol (methanol) when both substances are present.*] In: Alekseeva, M. V., ed. [*The determination of atmospheric pollutants*], Moscow, Medgiz, p. 187
7. Alekseeva, M. V. (1963) [*The separate determination of hydrochloric acid and chlorides.*] In: Alekseeva, M. V., ed. [*The determination of atmospheric pollutants*], Moscow, Medgiz, p. 31
8. Alekseeva, M. V., Andronov, B. E., Gurvic, S. S. & Zitkova, A. S. (1954) [*The determination of formaldehyde.*] In: [*The determination of harmful substances in the air of industrial areas*], Moscow, Goshimizdat, p. 360
9. Alekseeva, M. V. & Buštueva, K. A. (1954) *Gig. i Sanit.*, No. 4, p. 13
10. Alekseeva, M. V. & Rjazanov, V. A., ed. (1959) [*Determination of atmospheric pollutants*], Moscow, Medgiz
11. Alekseeva, M. V. & Samorodina, R. Ja. (1953) *Gig. i Sanit.*, No. 10, p. 42
12. Altshuller, A. P. & Bellar, T. (1963) *J. Air Pollut. Control Ass.*, **13,** 81
13. Altshuller, A. P. & Clemons, C. A. (1962) *Analyt. Chem.*, **34,** 466
14. Altshuller, A. P. & Lang, L. J. (1963) *Analyt. Chem.*, **35,** 1541
15. Altshuller, A. P. & McPherson, S. P. (1963) *J. Air Pollut. Control Ass.*, **13,** 109
16. American Conference of Governmental Industrial Hygienists (1958) *Manual of analytical methods*, Cincinnati, Ohio
17. American Conference of Governamental Industrial Hygienists (1962) *Air sampling instruments for evaluation of atmospheric contaminants*, 2nd ed., Cincinnati, Ohio
18. American Industrial Hygiene Association (1960) *Air pollution manual*, Part I, Detroit, Mich.
19. American Petroleum Institute (1959) *Sampling and analysis of waste gases and particulate matter.* In: *Manual on disposal of refining wastes*, 5th ed., New York
20. American Public Health Association (1955) *Standard methods for the examination of water, sewage and industrial wastes*, 10th ed., New York
21. American Society for Testing and Materials (1958) *Reference method for ammonia*, Philadelphia, Pa. (D1426)
22. American Society for Testing and Materials (1962) *ASTM standards on methods of atmospheric sampling and analysis*, Philadelphia, Pa.
23. Aruin, A. S. (1955) [*Determination of arsenious anhydride.*] In: Rjazanov, V. A., ed. [*Limits of allowable concentrations of atmospheric pollutants*, Book 2], Moscow, Medgiz, p. 126
24. Atkins, S. (1950) *Analyt. Chem.*, **22,** 947

25. Barnebey, H. L. & Davis, W. L. (1957) *J. Air Pollut. Control Ass.*, **7**, 86
26. Barnes, E. C. et al. (1949) *Analyt. Chem.*, **21**, 1281
27. Barnes, H. (1950) *Analyst*, **75**, 388
28. Batmanova, O. Y. (1951) *Gig. i Sanit.*, No. 12, 17
29. Belcher, R., Leonard, M. A. & West, T. S. (1959) *J. chem. Soc.*, p. 3577
30. Belcher, R. & West, T. S. (1961) *Talanta*, **5**, 863
31. Bellar, T. et al. (1962) *Analyt. Chem.*, **34**, 768
32. Bertolocino, R. J. & Barney, J. E. (1957) *Analyt. Chem.*, **29**, 281
33. Besson, A. & Pelletier, J. (1960) *Techn. Sci. munic.*, **55**, 90
34. Bokhoven, C. & Thomassen, P. H. (1961) *Nature (Lond.)*, **190**, 435
35. Bowen, E. J. & Brocklehurst, B. (1955) *J. chem. Soc.*, p. 4320
36. Brash, M. P. (1960) *Appl. Spectrosc.*, **14**, 2
37. Braverman, M. M., Hochheiser, S. & Jacobs, M. B. (1957) *Amer. Industr. Hyg. Ass. Quart.*, **18**, 132
38. Brewer, A. W. & Milford, J. R. (1960) *Proc. roy. soc. b*, **256**, 470
39. British Standards Institution (1951) *Deposit gauges for atmospheric pollution*, London (BS 1747)
40. Bryan, R. J. & Romanovsky, J. C. (1956) *Instrums Automn*, **29**, 2432
41. Buck, M. (1963) *Z. analyt. Chem.*, **193**, 101
42. Budd, M. S. & Bezick, H. A. (1952) *Analyt. Chem.*, **24**, 1536
43. Byers, D. H. & Saltzman, B. E. (1958) *Amer. industr. Hyg. Ass. J.*, **19**, 251
44. Cadle, R. D. et al. (1956) *Sampli.g procedures.* In: Magill, P. L., Holden, F. R. & Ackley, C., ed., *Air pollution handbook*, New York, McGraw-Hill
45. Caldwell, J. R. & Moyer, H. V. (1935) *Industr. Engng Chem. analyt. Edn*, **7**, 38
46. California State Department of Public Health (1960) *Recommended methods in air pollution measurements*, Berkeley, Calif. (Publication no. 1-20)
47. Cambi. F. (1961) *Sampling, analysis and instrumentation in the field of air pollution*, In: *Air pollution*, Geneva *(World Health Organization: Monograph Series*, No. 46), p. 63
48. Chaiken, S. W., Parks, T. S. & Glassbrook, C. (1956) US Patent no. 2,741.544
49. Chambers, L. A. (1956) In: Yaffe, C. D., Byers, D. H. & Hosey, A. D., ed., *Encyclopedia of instrumentation for industrial hygiene*, Ann Arbor, Mich., University of Michigan
50. Cholak, J. (1956) *Analytical methods.* In: Magill, P. L., Holden, F. R. & Ackley, C., ed., *Air pollution handbook*, New York, McGraw-Hill (section 11)
51. Cholak, J. & Hubbard, D. M. (1948) *Analyt. Chem.*, **20**, 73
52. Cholak, J. et al. (1957) *Arch. environm.* Hlth, **15**, 198
53. Cleary, G. J. (1962) *J. Chromat.*, **9**, 204
54. Clifton, Marjorie (1964) *Proc. roy Soc. Med.*, **57**, 1013
55. Coates, V. J. & Anacreon, R. (1957) *Improving the measurement of weak infrared absorption bands using expanded-scale, double-beam spectrophotometry.* In: *Proceedings of the conference on analytical chemistry and applied spectroscopy, Pittsburgh, Pa., March* 1957, Washington, D.C., American Chemical Society
56. Cohen, I. R. & Altshuller, A. P. (1961) *Analyt. Chem.*, **33**, 726
57. Cole, A. F. W. & Katz, M. (1966) *J. Air Pollut. Control Ass.*, **16**, 201
58. Commins, B. T. (1958) *Analyst*, **83**, 386
59. Commins, B. T. (1958) *Int. J. Air Pollut.*, **1**, 14
60. Commins, B. T. (1962) *Interim report on the study of techniques for determination of polycyclic aromatic hydrocarbons in air* In: Sawicki, E. & Cassel, K. Jr., ed. *Analysis of carcinogenic air pollutants* [Symposium on the Analysis of Carcinogenic Air Pollutants, Cincinnati, Ohio, 1961]. Washington, D.C., US Public Health Service (National Cancer Institute Monograph No. 9)
61. Commins, B. T. (1963) *Analyst*, **88**, 364

62. Cooper, R. L. (1954) *Analyst, Lond.*, **79,** 573
63. Craxford, S. R., Slimming, D. W. & Wilkins, E. T. (1960) *Measurement of air pollution.* In: *Proceedings of the 27th annual conference of the National Society for Clean Air, Harrogate*, London
64. Crocker, W., Zimmerman, P. W. & Hitchcock, A. E. (1932) *Contr. Boyce Thompson Inst. Pl. Res.*, **4,** 177
65. Cummings, W. G. & Redfearn, M. W. (1957) *J. Inst. Fuel*, *30*, 628
66. Cvejanovic, G. J. (1962) *Analyt. Chem.*, **34,** 654
76. Darley, E. F. (1966) *J. Air Pollut. Control Ass.*, **16,** 145
68. Darley, E. F. et al. (1963) *Arch. environm. Hlth*, **6,** 761
69. Davies, C. N. (1952) *Proc. Instn. mech. Engrs (B)*, **185,** 199
70. Dikun, P. P. (1961) *Vop. Onkol.*, **6,** No. 11, p. 75
71. Drinker, P. & Hatch, T. (1954) *Industrial dust*, New York, McGraw-Hill, p. 89
72. Dubois, L., Corkery, A. & Monkman, J. H. (1960) *Int. J. Air Pollut.*, **2,** 236
73. Dubois, L. & Monkman, J. L. (1961) *Amer. Industr. Hgy. Ass. J.*, **22,** 4
74. Dubois, L., Zdrojewski, A. & Monkman, J. L. (1966) *J. Air Pollut. Control Ass.* **16,** 135
75. Dubois, L. et al. (1964) *J. Air Pollut. Control Ass.*, **14,** 295
76. Efremova, G. P. (1955) [*Determination of phenol.*] In: Rjazanov, V. A., ed, [*Limits of allowable concentrations of atmospheric pollutants*, Book 2], Moscow. Medgiz, p. 99
77. Elkins, H. B. & Ruotolo, B. P. W. (1953) *Amer. industr. Hyg. Ass. Quart.*, **14,** 111
78. Epstein, P. S. (1929) *Z. Phys.*, **54,** 537
79. Fenton, A. F. (1964) *Ir. Nat. J.*, **14,** 237
80. Foran, M. R., Gibbons, E. V. & Wellington, J. R. (1958) *Chemistry in Canada*, **10,** 33
81. Fritz, J. S. & Yamamura, S. S. (1955) *Analyt. Chem.* **27** 1461
82. Gisclard, J. B. (1945) *Industr. Engng Chem. analyt. Edn*, **17,** 196
83. Gluckauf, E. et al. (1944) *J. chem. Soc.*, **1,** 1
84. Goetz, A. (1953) *Amer. J. publ. Hlth*, **43,** 150
85. Goetz, A. & Kallai, T. (1962) *J. Air Pollut. Control Ass.*, **13,** 437
86. Goetz, A., Stevenson, H. J. R. & Preining, O. (1960) *J. Air Pollut. Control Ass.*, **10,** 378
87. Gottschlich, C. F. (1962) In: Stern, A. C., ed., *Air pollution*, vol. II, New York, Academic Press, p. 30
88. Grant, G. A., Katz, M. & Haines, R. W. (1951) *Canad. J. Technol.*, **29,** 43
89. Grant, W. M. (1947) *Analyt. Chem.*, **19,** 345
90. Great Britain, Department of Scientific and Industrial Research (1932) *The investigation of atmospheric pollution*, London, H. M. Stationery Office
91. Great Britain, Department of Scientific and Industrial Research (1939) *Methods for the detection of toxic gases in industry*, London, H. M. Stationery Office
92. Great Britain, Department of Scientific and Industrial Research (1948) *Measurement of sulphur dioxide with the lead peroxide instrument*, London, H. M, Stationery Office
93. Green, H. L. & Lane, W. R. (1957) *Particulate clouds, dusts, smokes and mists*, London, Spon
94. Greenhalgh, R. & Riley, I. P. (1962) *J. Analyt. Chem.*, **189,** 431
95. Grever, P. M. & Cook, W. A. (1960) *Arch. industr. Hyg.*, **21,** 233
96. Gruber, C. W. & Alpaugh, E. L. (1954) *J. Air Pollut. Control Ass.*, **4,** 143
97. Gustafsson. L. (1960) *Talanta*, **4,** 227
98. Haagen-Smit, A. J. & Brunelle, M. F. (1958) *Int. J. Air Pollut.*, **1,** 51
99. Hall, S. R. (1952) *Analyt. Chem.*, **24,** 996
100. Hanker, J. S., Gelberg, A. & Witten, B. (1958) *Analyt. Chem.*, **30,** 93

101. Hauser, T. R. & Cummins, R. L. (1964) *Analyt. Chem.*, **36**, 679
102. Helwig, H. L. & Gordon, C. L. (1959) *Analyt. Chem.*, **30**, 1810
103. Hemeon, W. C. N., Haines, G. F. & Ide, H. M. (1953) *J. Air Pollut. Control Ass.*, **3**, 22
104. Hendrickson, E. F. (1962) *Air Sampling.* In: Stern, A. C., ed., *Air pollution*, vol. 1, New York, Academic Press
105. Hersch, P. & Sambucetti, C. J. (1963) *Galvanic-coulometric monitoring of carbon monoxide.* In: *Proceedings of the Pittsburgh conference on analytical chemistry and applied spectroscopy, March 1963*, Washington, D.C., American Chemical Society
106. Hochheiser, S. (1964) *Methods of measuring and monitoring atmospheric sulfur dioxide*, Washington, D.C., US Public Health Service, Division of Air Pollution (Publication No. 999-AP-6)
107. Houghton, G. U. & Pelley, R. G. (1937) *Analyst*, **62**, 117
108. Hritinina, A. (1940) reproduced in: Alekseeva, M. V., ed. [*The determination of atmospheric pollutants*], Moscow, Medgiz, p. 66
109. Hrustaleva, V. A. (1963) In: Alekseeva, M.V., ed. [*The determination of atmospheric pollutants*], Moscow, Medgiz, p. 193
110. Huitt, A. A. & Lodge, J. P. (1964) *Analyt. Chem.*, **36**, 1305
111. International Joint Commission (USA & Canada), Technical Advisory Board (1959) *Report on air pollution in the Detroit-Windsor area*, Washington, D.C. & Ottawa, Canada
112. Ivie, J. O. et al. (1963) *J. Air Pollut. Control Ass.*, **13**, 355
113. Jacobs, M. B. (1949) *The analytical chemistry of industrial poisons, hazards and solvents*, New York & London, Interscience
114. Jacobs, M. B. (1959) *Techniques for measurement of hydrogen sulfide and sulfur oxides.* In: Lodge, J. P., ed., *Atmospheric chemistry of chlorine and sulfur compounds*, Washington, D.C., American Geophysical Union
115. Jacobs, M. B. (1960) *The chemical analysis of air pollutants*, New York & London, Interscience
116. Jacobs, M. B., Braverman, M. M. & Hochheiser, S. (1957) *Analyt. Chem.*, **29**, 1349
117. Jacobs, M. B. & Greenburg, L. (1956) *Industr. Engng Chem.*, **48**, 1517
118. Jacobs, M. B. & Hochheiser, S. (1958) *Analyt. Chem.*, **30**, 426
119. Jeffrey, P. G. & Williams, D. (1961) *Analyst*, **86**, 590
120. Kaiser, E. R. (1962) *Odor and its measurement.* In: Stern, A. C., ed., *Air pollution*, vol. 1, New York, Academic Press
121. Kanno, S. (1959) *J. Air Pollut. Control Ass.*, **1**, 231
122. Katz, M. (1949) *Industr. Engng Chem.*, **41**, 2450
123. Katz, M. (1950) *Analyt. Chem.*, **22**, 1040
124. Katz, M. (1952) *The photoelectric determination of atmospheric sulfur dioxide by dilute starch-iodine solutions* In: *Air pollution. Proceedings of the United States Technical Conference on Air Pollution* [*Washington, D.C., 1950*] New York, Toronto, London, McGraw-Hill, p. 580
125. Katz, M. (1961) *Some aspects of the physical and chemical nature of air pollution.* In: *Air pollution*, Geneva *(World Health Organization: Monograph Series*, No. 46) p. 97
126. Katz, M. & Clayton, G. D. (1953) *Proc. Amer. Soc. Test. Mater.*, **53**, 1136
127. Katz, M. & Sanderson, H. P. (1958) *Spec. techn. Publs Amer. Soc. Test. Mater.*, No. 250, p. 29
128. Katz, M., Sanderson, H. P. & Ferguson, M. B. (1958) *Analyt. Chem.*, 30, 1172
129. Katz, M. et al. (1939) *Effect of sulfur dioxide on vegetation*, Ottawa, National Research Council of Canada (Publication No. 815)
130. Keagy, D. M. et al. (1961) *J. Air Pollut. Control Ass.*, **11**, 270

131. Keenan, R. G. & Faishall, L. T. (1944) *J. industr. Hyg.*, **26,** 241
132. Keenan, R. G. et al. (1963) *J. Amer. industr. Hyg. Ass.*, **24,** 481
133. Keily, H. J. & Rodgers, L. B. (1955) *Analyt. Chem.*, **31,** 2052
134. Kemény, E. (1962) *J. Air Pollut. Control Ass.*, **12,** 278
135. Kethley, T. W., Gordon, M. T. & Orr, C. O. (1952) *Science*, **116,** 368
136. Kitagawa, T. (1960) *The rapid measurement of toxic gases and vapours.* In: *Proceedings of the XIII international congress on Occupational health, New York . . . July, 1960*
137. Kolthoff, I. M. & Stenger, V. A. (1947) *Volumetric analysis*, vol. 2, New York, Interscience
138. Krivoručko, F. D. (1953) *Gig. i Sanit.*, No. 3, p. 53
139. Kusnetz, H. L. (1962) *Inertial collectors.* In: American Conference of Governmental Industrial Hygienists, *Air sampling instruments for evaluation of atmospheric contaminants*, 2nd ed., Cincinnati, Ohio
140. Landis, F. P. & Coons, M. C. (1954) *Appl. Spectrosc.*, **8,** 71
141. Langer, G., Pierrard, J. & Yamate, G. (1964) *Int. J. Air Wat. Pollut.*, **8,** 167
142. Leighton, P. A. (1961) *Photochemistry of air pollution*, New York & London, Academic Press
143. Levaggi, D. A. & Feldstein, M. (1964) *Amer. industr. Hyg. Ass. J.*, **25,** 64
144. Liederman, D., Bowen, J. E. & Milner, O. I. (1959) *Analyt. Chem.*, **31,** 2052
145. Littman, F. E. & Benoliel, R. W. (1953) *Analyt. Chem.*, **25,** 1480
146. Littman, F. E. & Denton, J. Q. (1956) *Analyt. Chem.*, **28,** 945
147. Lodge, J. P. (1956) *J. Met.*, **13,** 406
148. Loginov, N. E. (1931) *Bjull. Leningrad. Inst. Gig. Tr. Teh. Besopast.*, No. 7–8, p. 42
149. Macdowall, F. D. H. et al. (1964) *Can. J. Pl. Sci.*, **44,** 410
150. Mader, P. P., Hamming, W. J. & Bellin, A. (1950) *Analyt. Chem.*, **22,** 1181
151. Madison, J. J. (1958) *Analyt. Chem.*, **30,** 1859
152. Magill, P. L. & Cralley, L. V. (1954) In: *Air pollution abatement manual*, Washington, D.C., Manufacturing Chemists Association
153. Manufacturing Chemists Association (1954) *Air pollution abatement manual*, Washington D.C.
154. Middleton, J. T., Emik, L. O. & Taylor, O. C. (1965) *J. Air Pollut. Control Ass.*, **15**, 476
155. Middleton, J. T. & Paulus, A. O. (1956) *Arch. environm. Hlth*, **14,** 526
156. Moore, G. E., Cole, A. F. W. & Katz, M. (1957) *J. Air Pollut. Control Ass.*, **7,** 25
157. Moore, G. E. & Katz, M. (1960) *Int. J. Air Pollut.*, **2,** 221 ;
158. Morris, R. A. & Chapman, R. L. (1961) *J. Air Pollut. Control Ass.*, **11,** 467
159. Moureu, H. (1964) *Proc. roy. Soc. Med.*, **57,** 1015
160. Muel, B. & Lacroix, G. (1960) *Bull. Soc. chim. Fr.*, p. 2139
161. Nader, J. A. (1959) *J. Air Pollut. Control Ass.*, **9,** 59
162. Nash, T. (1961) *J. sci. Instrum.*, **38,** 480
163. Nash, T. (1964) *Int. J. Air Wat. Pollut.*, **8,** 121
164. Nauman, R. V. et al. (1960) *Analyt. Chem.*, **32,** 1307
165. Neuberger, H., Panofsky, H. & Sekera, Z. (1956) *Physics of the atmosphere.* In: Magill, P. L., Holden, F. R. & Ackley, C., ed., *Air pollution handbook*, New York, McGraw-Hill
166. O'Konski, C. T. & Doyle, G. F. (1955) *Analyt. Chem.*, **27,** 694
167. Organization for Economic Co-operation and Development, Working Party on Methods of Measuring Air Pollution . . . (1964) *Methods of measuring air pollution*, Paris
168. Owen, L. E. et al. (1951) *J. Amer. industr. Hyg. Ass.*, **12,** 112
169. Panin, K. P. (1959) *Gig. i Sanit.*, No. 9, p. 70
170. Pate, J. B., Lodge, J. P. & Wartburg, A. F. (1962) *Analyt. Chem.*, **34,** 1660

171. Pate, J. B. & Tabor, E. C. (1962) *Amer. industr. Hyg. Ass. J.*, **23,** 145
172. Paul', E. I. & Vinogradova, V. A. (1960) *Gig. Tr. prof. Zabol.*, No. 2, p. 53
173. Paulus, H. J., Floyd, E. P. & Byers, D. H. (1954) *Amer. industr. Hyg. Ass. Quart.*, **15**, 4
174. Paulus, H. J., Floyd, E. P. & Byers, D. H. (1954) *Amer. industr. Hyg. Ass. Quart.*, **15,** 277
175. Pavlu, J. & Sula, J. (1960) *Colln Czech. chem. Commun. Engl. Edn*, **25,** 2461
176. Pelletier, J. (1963) *Int. J. Air Wat. Pollut.*, **7,** 993
177. Personov, R. I. (1962) *Z. analit. Him.*, **17,** 506
178. Peterson, G. E. et al. (1950) *Analyt. Chem.*, **22,** 1197
179. Petrilli, F. L. (1962) *Bull. Wld. Hlth Org.*, **26,** 495
180. Polezaev, N. G. (1937) *Gig. Tr. Teh. Bezopast.*, No. 6, p. 38
181. Polezaev, N. G. (1952) [*Determination of lead aerosol.*] In: Rjazanov, V. A., ed. [*Limits of allowable concentrations of atmospheric pollutants*, Book 1], Moscow, Medgiz, p. 140
182. Porter, K. & Volman, D. H. (1962) *Analyt. Chem.*, **34,** 748
183. Preining, O. (1962) *Staub*, **22,** 129
184. Ranz, W. E. & Wong, J. B. (1952) *Ind. Engng Chem.*, **44,** 1371
185. Raynor, G. S. (1957) *J. Air Pollut. Control Ass.*, **7,** 122
186. Regener, V. H. (1959) *Automatic chemical determination of atmospheric ozone.* In: *Advances in chemistry*, ser. 21, Washington, D.C., American Chemical Society
187. Regener, V. H. (1964) *J. geophys. Res.*, **69,** 3795
188. Renzetti, N. A., ed. (1955) *An aerometric survey of the Los Angeles basin, August-November 1954*, San Marino, Calif., Air Pollution Foundation (Report No. 9)
189. Renzetti, N. A. (1957) *Analyt. Chem.*, **29,** 869
190. Renzetti, N. A. (1962) *Automatic instrumentation.* In: Stern, A. C., ed., *Air pollution*, vol. 1, New York, Academic Press
191. Renzetti, N. A. & Romanovsky, J. C. (1956) *J. Air Pollut. Control Ass.*, **6,** 154
192. Ripley, D. W., Clingenpeel, J. M. & Hurn, R. W. (1964) *Int. J. Air Wat. Pollut.*, **8,** 455
193. Rjazanov, V. A., ed. (1952-60) [*Limits of allowable concentrations of atmospheric pollutants*, Books 1-4], Moscow, Medgiz
194. Rodebush, W. H. (1950) *General properties of aerosols.* In: US Atomic Energy Commission, *Handbook on aerosols*, Washington, D.C.
195. Rogers, L. H. (1957) *Chem. Engng Prog.*, **53,** 381
196. Rondia, D. (1962) *Int. J. Air Wat. Pollut.*, **6,** 353
197. Rose, H. E. & Wood, A. J. (1956) *An introduction to electrostatic precipitation in theory and practice*, London, Constable
198. Rosenblatt, P. & LaMer, V. K. (1946) *Physiol. Rev.*, **70,** 385
199. Rossano, A. T. (1962) *The air pollution survey.* In: Stern, A. C., ed. *Air pollution*, vol. 1, New York, Academic Press
200. Roubal, J., Sdrazil, J. & Picha, F. (1954) *Čs. Hyg.*, **3,** 188
201. Saltzman, B. E. (1953) *Analyt. Chem.*, **25,** 493
202. Saltzman, B. E. (1954) *Analyt. Chem.*, **26,** 1949
203. Saltzman, B. E. (1961) *Analyt. Chem.*, **33,** 1100
204. Saltzman, B. E. (1962) *Amer. industr. Hyg. Ass. J.*, **23,** 112
205. Saltzman, B. E. (1962) *Direct reading colorimetric indicators.* In: American Conference of Governmental Industrial Hygienists*Air sampling instruments for evaluation of atmospheric contaminants*, 2nd ed., Cincinnati, Ohio
206. Saltzman, B. E. (1965) *Selected methods for the measurement of air pollutants*, Cincinnati, Ohio, R. A. Taft Sanitary Engineering Centre, Division of Air Pollution, Interbranch Chemical Advisory Committee (Publication No. 999-AP-11)

207. Saltzman, B. E. & Gilbert, N. (1959) *Amer. industr. Hyg. Ass. J.*, **20**, 379; *Analyt. Chem.*, **31**, 1914
208. Saltzman, B. E. & Wartburg, A. F. (1965) *Analyt. Chem.*, **37**, 779
209. Sandell, E. B. (1950) *Colorimetric determination of traces of metals*, New York, Interscience
210. Sanderson, H. P., Bradt, P. & Katz, M. (1963) *J. Air Pollut. Control Ass.*, **13**, 461
211. Sanderson, H. P. & Katz, M. (1963) *J. Air Pollut. Control Ass.*, **13**, 476
212. Sanderson, H. P., Penner, P. E. & Katz, M. (1964) *Analyt. Chem.*, **36**, 2296
213. Sawicki, E. (1964) *Chemist Analyst*, **53**, 24, 56, 88
214. Sawicki, E. & Cassel, K. Jr., ed. (1962) *Analysis of carcinogenic air pollutants* [Symposium on the Analysis of Carcinogenic Air Pollutants, Cincinnati, Ohio, 1961], Washington, D.C., US Public Health Service (National Cancer Institute Monograph No. 9)
215. Sawicki, E. et al. (1960) *J. Amer. industr. Hyg. Ass.*, **21**, 443
216. Sawicki, E. et al. (1961) *Analyt. Chem.*, **33**, 93
217. Sawicki, E. et al. (1964) *Chemist Analyst*, **53**, 6
218. Sawyer, K. F. & Walton, W. H. (1950) *J. scient. Instrum.*, **27**, 272
219. Sawyer, R. R., Behnke, F. W. & Adams, N. I. (1958) *The infrared analysis of trace gases using long path gas cell and ordinate expansion.* In: *Proceedings of the 51st annual meeting of the Air Pollution Control Association, Philadelphia, Pa., June, 1958*, Pittsburgh, Pa., Air Pollution Control Association
220. Saxton, R. L. & Ranz, W. E. (1952) *J. appl. Physiol.*, **23**, 917
221. Schadt, C. et al. (1950) *Arch. industr. Hyg.*, **1**, 556
222. Senderihina, D. P. (1955) *Inf.-metod. Mater. Inst. F. F. Erisman.*, No. 5, 21
223. Shaffer, P. A., Briglio, A. & Brockman, J. A. (1948) *Analyt. Chem.*, **20**, 1008
224. Shaw, N. & Owens, J. S. (1925) *The smoke problem of great cities*, London, Constable
225. Shepherd, M. (1947) *Analyt. Chem.*, **19**, 77
226. Shepherd, M. et al. (1951) *J. Analyt. Chem., Wash.*, **23**, 1431
227. Sinclair, D. (1950) *Optical properties of aerosols.* In: US Atomic Energy Commission, *Handbook on aerosols*, Washington, D.C.
228. Sinclair, D. (1953) *J. Air Pollut. Control Ass.*, **3**, 51
229. Skeggs, L. T. (1957) *Amer. J. clin. Path.*, **28**, 311
230. Skye, E. (1965) *Acta Phytogeogr. Suec.*, **50**, 285
231. Smith, R. B. & Fries, B. S. T. (1931) *J. industr. Hyg.*, **13**, 338
232. Smith, R. G. & Diamond. P. (1952) *Amer. industr. Hyg. Ass. Quart.*, **13**, 235
233. Smith, R. G., MacEwen, J. D. & Barrow, R. E. (1959) *J. Amer. industr. Hyg. Ass.*, **20**, 142
234. Smith, R. G. et al. (1952) *Analyt. Chem.*, **24**, 406
235. Smith, W. J. & Surprenant, N. F. (1953) *Proc. Am. Soc. Test. Mater.*, **53**, 1122
236. Stalker, W. W., Dickerson, R. C. & Kramer, G. C. (1963) *J. Amer. industr. Hyg. Ass.*, **24**, 68
237. Stang, A. M., Zatek, J. E. & Robson, C. D. (1951) *Amer. industr. Hyg. Ass. Quart.*, **12**, 5
238. Steigman, A. (1942) *J. Soc. chem. Industr.*, **61**, 18
239. Steigman, A. (1950) *Analyt. Chem.*, **22**, 492
240. Stephens, E. R. et al. (1956) *Industr. Engng Chem.*, **48**, 1498
241. Still, C. W., Willis, C. W. & Willis, C. P. (1959) *Analyt. Chem.*, **31**, 598
242. Strange, J. P. (1957) *Analyt. Chem.*, **29**, 1878
243. Stratmann, H. (1954) *Mikrochim. Acta*, **6**, 668
244. Sullivan, J. L. (1962) *J. Air Pollut. Control Ass.*, **12**, 474
245. Terraglio, F. P. & Manganelli, R. M. (1962) *Analyt. Chem.*, **34**, 675
246. Thomas, M. D. (1952) In: *Proceedings of the II National Air Pollution Symposium*, Los Angeles, Calif., Stanford Research Institute, p. 16

247. Thomas, M. D. (1961) *Effects of air pollution on plants.* In: *Air pollution*, Geneva (*World Health Organization:* Monograph Series, No. 46)
248. Thomas, M. D. (1962) *Int. J. Air Wat. Pollut.*, **6**, 443
249. Thomas, M. D. & Abersold, J. N. (1929) *Industr. Engng Chem. analyt. edn*, **1**, 14
250. Thomas, M. D. & Abersold, J. N. (1932) *Industr. Engng Chem. analyt. Edn*, **4**, 253
251. Thomas, M. D. & Cross, R. J. (1928) *Industr. Engng Chem.*, **20**, 645
252. Thomas, F. W. & Davidson, C. M. (1961) *J. Air Pollut. Control Ass.*, **11**, 24
253. Thomas, M. D., St John, G. A. & Chaiken, S. W. (1958) *Spec. techn Publs Amer. Soc. Test. Mater.*, No. 250, p. 49
254. Thomas, M. D. et al. (1943) *Industr. Engng Chem. analyt. Edn*, **15**, 287
255. Thomas, M. D. et al. (1946) *Industr. Engng Chem. analyt. Edn*, **18**, 383
256. Thomas, M. D. et al. (1956) *Analyt. Chem.*, **28**, 1810
257. US Public Health Service (1958) *The air over Louisville: summary of a joint report*, Cincinnati, Ohio, R. A. Taft Sanitary Engineering Center
258. Urone, P. F. & Boggs, W. E. (1957) *Analyt. Chem.*, **23**, 1517
259. Vasak, V. & Sedivec, V. (1952) *Chemické Listy*, **46**, 341
260. Vassy, A. (1961) *C. R. Acad. Sci. (Paris)*, **252**, 571
261. Walford, G. A. & Harley, A. J. (1952) *Amer. industr. Hyg. Ass. Quart.*, **13**, 232
262. Walkley, Janet (1959) *J. Amer. industr. Hyg. Ass.*, **20**, 241
263. Waller, R. E. (1963) *Int. J. Air Wat. Pollut.*, **7**, 773
264. Wartburg, A. F., Brewer, A. W. & Lodge, J. P. (1964) *Int. J. Air Wat. Pollut.*, **8**, 21
265. Washburn, H. W. & Austin, R. R. (1952) *The continuous measurement of SO_2 and H_2S concentrations by automatic titration.* In: *Air pollution. Proceedings of the United States Technical Conference on Air Pollution* [Washington, D.C., 1950], New York, Toronto, London, McGraw-Hill
266. Waters, J. L. & Hartz, N. W. (1952) *Instruments*, **25**, 622
267. Watson, H. H. & Hounam, R. F. (1948) *Colliery Guard.*, **176**, 447
268. Weber, C. W. & Howard, O. H. (1959) *Arch. environm. Hlth*, **19**, 355
269. Weisz, H. (1961) *Microanalysis by the ring oven technique*, New York, London & Paris, Pergamon
270. Welch, A. F. & Terry, J. P. (1960) *J. Amer. industr. Hyg. Ass.*, **21**, 316
271. West, P. W. (1966) *J. Air Pollut. Control Ass.*, **16**, 601
272. West, P. W. & Gaeke, G. C. (1956) *Analyt. Chem.*, **28**, 1816
273. West, P. W. & Ordoveza, F. E. (1962) *Analyt. Chem.*, **34**, 1324
274. West, P. W. et al. (1960) *Analyt. Chem.*, **32**, 943
275. Wilson, W. L. (1954) *J. Air Pollut. Control Ass.*, **4**, 8
276. Yaffe, C. D., Byers, D. H. & Hosey, A. D., ed. (1956) *Encyclopedia of instrumentation for industrial hygiene*, Ann Arbor, Mich., University of Michigan
277. Yocum, J. E. et al. (1956) *The effect of hydrogen chloride on the operation of the Thomas autometer.* In: *Proceedings of the 49th annual meeting of the Air Pollution Control Association, Buffalo, N.Y.*, Pittsburgh, Pa., Air Pollution Control Association
278. Zaleiko, N. S. (1963) *J. Air Pollut. Control Ass.*, **13**, 531
279. Zeidberg, L. D. et al. (1961) *J. Air Pollut. Control Ass.*, **11**, 289
280. Zimmerman, P. W. (1952) *Effects on plants of impurities associated with air pollution.* In: *Air pollution. Proceedings of the United States Technical Conference on Air Pollution* [Washington, D.C., 1950], New York, Toronto, London, McGraw-Hill
281. Zurlo, N. & Griffini, A. M. (1962) *Med. d. Lavoro*, **53**, 325
282. Zurlo, N. & Metrico, L. (1962) *Metodi semplici di microdeterminazione dei tossici industriali dispersi nell'aria*, Milan, Clinica del Lavoro

244. Thomas, M. D. (1961) Effects of air pollution on plants. In Air pollution. World Health Organization Monograph Series, No. 46, [illegible]
245. Thomas, M. D. (1962) [illegible]
246. [illegible]
247. [illegible]
248. [illegible]
249. [illegible]
250. [illegible]
251. [illegible]
252. [illegible]
253. [illegible]
254. [illegible]
255. [illegible]
256. [illegible]
257. U.S. Public Health Service (1969) Air quality criteria for [illegible] Cincinnati, Ohio. [illegible]
258. [illegible]
259. [illegible]
260. [illegible]
261. [illegible]
262. [illegible]
263. [illegible]
264. [illegible]
265. [illegible]
266. [illegible]
267. [illegible]
268. [illegible]
269. [illegible]
270. [illegible]
271. [illegible]
272. [illegible]
273. [illegible]
274. [illegible]
275. [illegible]
276. [illegible]
277. [illegible]
278. [illegible]
279. [illegible]
280. Zimmerman, P. W. [illegible]
281. [illegible]
282. [illegible]